NANTAHALA
NATIONAL FOREST
A History

MARCI SPENCER

Foreword by George Ellison
A Commentary by James Lewis, PhD

Published by The History Press
Charleston, SC
www.historypress.net

First published 2017

Manufactured in the United States

ISBN 9781467136372

Library of Congress Control Number: 2017938354

Notice: The information in this book is true and complete to the best of our knowledge. It is offered without guarantee on the part of the author or The History Press. The author and The History Press disclaim all liability in connection with the use of this book.

To my mother,
Alice Leona Creasman Lance,
who was ready, always, for a new adventure.

"Let's go!"

CONTENTS

Contents

FOREWORD

Let the drama begin," Marci Spencer proclaims at the conclusion of her introductory chapter for this wonderfully written account of the political, cultural and natural history of the Nantahala National Forest (NNF). "Drama" is not an inappropriate way to describe this rendering of diverse storylines played out in an amphitheater of 500,000 acres in the far southwestern tip of North Carolina.

That vast setting is held in place today not only by highways and byways but also by a plethora of trails and rivers that Spencer followed as she tracked down the next story, the next informant, the next wild place that awaited her just over the next ridge or the one just beyond that one. The names of not a few of the shining rivers—Tuckasegee, Nantahala, Oconoluftee, Chattooga, Cullasaja—were remindful as well of a time when this land was the homeland of the ancient Cherokees. And the Bartram Trail and Appalachian Trail (A.T.) wend their way through each chapter, connecting people and places becoming, as it were, major "characters" in the narrative.

Anyone familiar with Spencer's previous books about Clingmans Dome and Pisgah National Forest is well aware of her ability to assemble and digest archival, scholarly or journalistic materials. But she is perhaps most comfortable—and at her best—when walking a trail with, say, a Forest Service ranger or a botanist, listening carefully to what they have to say.

As a writer who has resided in and written about the people and places of this region for going on half a century, I can sense how exciting it was for Spencer to plan and then execute her fact-gathering ventures into every nook and cranny within each of the three ranger districts. But I must admit that I had supposed I knew all the stories about the region and had already written something about most of them. Wrong. I see now that I had only scratched the surface.

Her cast of characters is composed of just about every figure of significance in the region's history, from plant-hunter William Bartram, author of *Bartram's Travels* (1791), which influenced the tradition of descriptive literature in this country as well as the development of Romantic literature in England; to Cherokee chief Attakullakulla, who had, as a younger man, been to England to meet the king and in 1776 encountered Bartram along what is now known as Bartram's Trail, seventy-some miles of which traverse NNF lands; to Burt Kornegay, an outdoor guide himself, who has served as founder, cartographer and president of the NC Bartram Trail Society; to hiking trail advocate Reverend A. Rufus Morgan, who founded the Nantahala Hiking Club and has a trail named for him within the NNF but is perhaps best remembered for having walked the rugged trail in the Great Smokies that accesses Mount Le Conte 172 times; and on to my great friend the indomitable Sally Kesler, who served as Reverend Morgan's "right hand man" for many years as they walked the trails together and then as his "eyes" when he went blind but kept on walking.

Other "characters" in Spencer's "drama" that come readily to mind would be Thomas Griffiths, who came up the Indian Trail from Charleston, South Carolina, in 1776 and "collected nearly five tons of clay [kaolin] in Cowee Valley in 1776 to ship back to England's Wedgewood factories"; Nimrod Simpson Jarrett, a land speculator, ginseng trader and owner of talc and mica mines in the Nantahala Gorge whose property stretched from about one mile upstream of Nantahala's juncture with Jarretts Creek (named for him) to the current Ferebee Memorial Recreation Area in the Nantahala Gorge; Payson and Aurelia Kennedy and their four children, "who moved from Atlanta to the banks of the Nantahala River" in 1971 and established the Nantahala Outdoor Center, which in time has garnered a worldwide reputation as a whitewater recreational center, where many of this nation's Olympic-level paddlers train; Lewis Kearney, a former district ranger on the Nantahala National Forest, who after arriving in Western North Carolina in 1980 either initiated or helped implement an array of projects; Tsali, a Cherokee farmer who, along with several family members, was executed by the U.S. military for their resistance to forced removal to Oklahoma in 1838; Chief Junaluska, who was removed to Oklahoma but walked back home to Robbinsville and was awarded state citizenship, a 364-acre farm and $100 by the North Carolina government; Dr. J. Dan Pittillo, retired biology professor at Western Carolina University, who has through a long career identified most of the natural areas in WNC and stood up for them when threatened; Marshall McClung, retired USFS forester, who can search for a downed plane, fight a forest fire or slip on a costume and become Smokey Bear with equal aplomb; and Eric Rudolph, the thirty-six-year-old fugitive wanted by the Federal Bureau of Investigation (FBI) for bombing an abortion clinic in Alabama and the Olympic Park in Atlanta, killing two and injuring hundreds, who for five years hid out on NNF lands from authorities before being apprehended while sifting through a dumpster behind the Valley River Shopping Center in Murphy at 3:30 a.m. (so as to avoid coming face-to-face with his maimed victims, he took the cowardly way out by pleading guilty before being sentenced to four consecutive life sentences).

I don't know if you can tell a book by its cover. But I am certain that you can tell a book by its opening sentences. They need to reach out and grab you. No one will ever do better in that regard than Herman Melville, whose masterpiece, *Moby Dick*, opens this way: "Call me Ishmael."

Marci Spencer has a gift for crafting openings that assure the reader that what follows will be interesting and well written. Her first chapter tells, in part, the story of a multimillion-dollar drug heist involving a pilotless air plane that crashed on NNF land as an example of the sort of diverse problems USFS workers have to deal with on a regular basis:

> *"We heard a noise like an airplane flying real low," reported a fisherman to the sheriff of Macon County, "and then it went silent."*
>
> *"Then we heard a loud bang," his buddy added.*
>
> *"When did you hear this?" the sheriff asked.*
>
> *"Yesterday."*
>
> *"Yesterday?"*
>
> *"Yeah, we were fishing at Nantahala Lake. We fished all day and didn't get back home until after dark."*

Chapter 4, which contains much about the railway systems as they evolved in the Southern Appalachians, opens like this:

> *At the Asheville depot, Murphy-bound passengers boarded Murphy Branch rail line No. 17. Dubbed, in jest, the "Asheville Cannonball" for its ten- to fifteen-mile-per-hour speeds, the train crawled west. After passing stations at Candler and Canton and crossing the Pigeon River, it headed past Clyde*

and Waynesville toward one of its most dramatic challenges, 3,315-foot Balsam Gap. Engineers used double-headers, a second steam locomotive behind the first, to help climb steep grades.

Whether you are a native Western North Carolinian or a recent arrival, Marci Spencer's *Nantahala National Forest* will serve as a good read about the region's natural and human history and as a guide should you want to visit areas associated with specific events.

GEORGE ELLISON
Bryson City, North Carolina

A COMMENTARY

When Marci Spencer asked me to write an opening commentary for her latest book, I was curious, as a historian myself, to learn how it would differ from previous histories of the Nantahala National Forest. While historians can generally agree on facts, what they often differ on is the interpretation of those facts. And interpretation is often influenced by things like the author's motivation for writing, scholarly trends and who the intended audience is. When I revisited what those earlier histories had said about Nantahala, I realized that Marci, who also authored *Pisgah National Forest: A History*, is among what I would describe as the third wave of historians writing about national forests. How it differs from the early waves is important for understanding what Marci has achieved with *Nantahala National Forest*.

One point on which historians have long agreed: Since its establishment in 1905, the U.S. Forest Service has frequently found itself battling controversy. Furthermore, some controversies, such as the ones over clearcutting lumber or managing northern spotted owl habitat, have pushed the agency to change its philosophy of forest management, which in turn altered how it has managed the national forests. One of the oldest battles has been over who exactly is best suited to manage the nation's forested lands for the benefit of all—federal, state or local authorities, or even private citizens. From the time of its founding, the Forest Service has had to counter critics who argued that the federal government had no business managing lands. The Forest Service's first chief, Gifford Pinchot, promised at the outset that the agency would manage the national forests "for the greatest good of the greatest number in the long run," something individuals, corporations or state governments had not been willing or able to do. It would be a *national* forest service—although its forests initially were only in the West.

This promise to manage for the greatest good was immediately put to the test. By the time Congress had established the Forest Service, logging companies had been conducting large-scale operations in the Appalachian Mountains for more than two decades, clearcutting forests and then abandoning them instead of managing for the long term. Fires and flooding frequently resulted. Clearly the companies were unwilling to manage forests for the greatest—or even a greater—good. The Forest Service went on the offensive by supporting a new law that would empower the federal government to purchase private lands to protect watersheds, prevent forest fires and facilitate reforestation. In 1911, Congress passed a bill sponsored by Representative John Weeks of Massachusetts. And although most of the land to come under protection was in the East, national conservation groups supported the Weeks Act because conserving forests in the South and New England was in the nation's best interest. Among the first national forests created under the law was the Nantahala National Forest in 1920.

While the Weeks Act would eventually become a success story for the Forest Service, it was not at all clear in the 1920s that this would be the case. Establishment of the National Park Service in 1916 introduced direct competition over who would manage the country's more scenic forested landscapes. From early on, national parks were touted as "the people's playgrounds." But few people then—or even today—recognized that while national forests provide recreational opportunities similar to those of national parks—camping, hiking and fishing, to name just a few—they also permit natural resource extraction activities that includes logging, mining and grazing. The rivalry between the two agencies grew acrimonious over the next two decades, so much so that in the 1930s the U.S. Forest Service funded the research and writing of books documenting the history of individual national forests.

The Forest Service's decision to fund this first wave of national forest histories came at a critical time. Soon after Franklin Roosevelt became president in 1933, the agency found itself fighting to remain an independent entity. Secretary of the Interior Harold Ickes desperately wanted to have the Forest Service transferred to his department to combine with other federal land management bureaus, including the National Park Service, in a new Department of Natural Resources. Naturally, Ickes would be in charge. Thrown on the defensive and fighting for its survival, the Forest Service began publishing these histories in an effort to educate the public about national forests and why they mattered.

The first-wave books had several characteristics in common. They were primarily written to introduce readers to the many benefits national forests provide the people of the United States, most notably their contribution to the nation's wood supply and the varied offerings of recreational activities. Bearing nondescript titles such as *National Forests in the Southern Appalachians* and *Ouachita National Forest, Arkansas-Oklahoma*, the books are credited to the Forest Service instead of to individual authors. Not surprisingly, they have a decidedly promotional air about them, coming across more as travel guides than as history books. Targeting the summer vacationer, the books typically open with a very brief history of the featured forest before launching into an effusive description of its landscape and flora. Next, they tout the forest's proximity to major metropolitan areas. The agency hoped to persuade urban audiences in a car-crazy nation to make the one- or two-day trek to the forests "over comfortable roads," as one of the books counseled; further, the books reassured readers that, once there, most trails and campgrounds would be accessible by automobile. Accompanied by plenty of photos showing people enjoying the outdoors, the enticing descriptions of cool waters and shade trees were followed by listings of recreational offerings that detailed fishing and hunting areas and noted places to hike, camp and picnic.

Another commonality among these books is how they present the history itself. The narratives rush past the fact that Native Americans had occupied—and strategically manipulated—the land for centuries by the time European explorers and then settlers arrived. The assortment of residents of southern forests during the centuries immediately preceding the Forest Service's arrival on the scene do not fare much better. A few short sentences gloss over "four centuries of colorful history" (as *Florida's National Forests* characterizes it); the narrative then ushers readers along to the timber exploitation era of the late 1800s and early 1900s. Of course, without that exploitation there would be no need for the heroic arrival of the Forest Service, much less its subsequent land management derring-do. The books go on to include discussions of the Forest Service's management goals, informing readers that what foresters do will provide natural resources long into the future. Text boxes interspersed throughout the pages provide outdoor-recreation etiquette tips for urban visitors, such as reminders to extinguish campfires and report forest fires promptly.

About fifty years after these books came out, the Forest Service again found itself battling the Department of the Interior, which this time had allied itself with a private sector challenger. By the 1970s, the Sagebrush Rebellion had erupted across the American West. In hopes of exploiting forest resources with little oversight or interference, the movement's supporters demanded that federal lands be turned over to state or local control or sold to private owners. When they took office in 1981, President Reagan and his secretary of the interior, James Watt, openly supported these goals. Success for the Sagebrush Rebellion might very well have meant the Forest Service would cease to exist. That the attack originated in the executive branch, which oversees the agency and the Department of Agriculture in which it resides, made the situation seem very much like the 1930s all over again.

This time, however, the Forest Service was also under fire from environmental groups demanding the agency do a better job of managing forests and to slow, if not completely stop, their timber activities. The passage of environmental laws in the 1960s and 1970s—including the Wilderness Act (1964), the National Environmental Policy Act (1969) and National Forest Management Act (1976)—gave environmental activists a quiver full of legal arrows with which to attack the "timber beast," as they derisively referred to the Forest Service. Certainly, their lawsuits effectively wounded and slowed the Forest Service, leaving it largely unable to move forward with implementing its mandated forest management plans. It became not uncommon for a plan to take more than a decade to get through the approval process, by which time conditions on the land had changed so much that planners needed new assessments and analysis. Although things have improved for planners in the last decade, the problem persists.

As in the New Deal era, in the 1980s the agency turned to history to explain what the Forest Service and its foresters—as well as engineers and wildlife biologists and ecologists and others now involved in the planning process—did and why. This time, though, the Forest Service contracted historians from outside the agency to write scholarly accounts of individual forests and the nine forest service regions. This second wave of books, which included a summary history of Forest Service policies and descriptions of the natural and cultural resources within the agency's purview, might be viewed as a rear-guard action, one undertaken to win over a skeptical general public and executive branch. How effective it was in doing so is impossible to assess. But the historians did what they were hired to do: they produced scholarly, illustrated histories that traced the history of the land and its many occupants and managers, complete with footnotes. Published by the Forest Service, the books were frequently accessed by researchers but rarely perused by the reading public.

Now, more than thirty years after the second wave of histories began to appear, comes the third wave. Once again, the timing is significant. National forests—and federal lands in general—are under political pressure, having been targeted by activists who believe, like the Sagebrush Rebellion supporters a generation ago, that federal lands should be sold off or transferred to the states. Environmental groups continue to hamper the agency at times; however, concerted efforts between local environmental organizations and agency leaders at the national forest and district levels to communicate intentions and goals are leading to new levels of trust and cooperation.

It is worth noting that this third wave of history books is not funded by the Forest Service itself. Some are being published by university presses for academic audiences, while others come from publishers such as The History Press and are intended for general audiences.

Incorporating the best aspects of the first two waves, the third wave is characterized by lively history that encourages the reader to visit the forests and is built on a foundation of solid research. Some of the most accessible, pleasurable history about our national forests being produced today is by historians like Marci Spencer. Often a local resident possessing a passion for the place and the people about which they write, these historians have typically lived near the national forests they are writing about for years and are personally familiar with many of the key players and their roles in the national forest community. Their conversations with Forest Service personnel, the area residents who rely on the land for their livelihoods and visitors from near and far inform this third wave of history books—books the general public will actually read and will want to revisit again and again.

This knowledgeable approach and intelligent accessibility is just part of what Marci Spencer brings to *Nantahala National Forest*. And in *Nantahala*, as with her book *Pisgah National Forest*, Marci demonstrates a love for and knowledge of place that is substantial and intimate and engaging. After reading her books, you, too, will want to visit the places she describes—much like the feeling the books of the first wave induced for that generation of readers. And as the readers of the second wave did, you will come away with a deeper appreciation for and understanding of why for more than a century these lands have been worth protecting and conserving. By conveying that history in such relatable prose, she demonstrates what the history of the Nantahala and the other 153 national forests has made clear—that, although imperfect, the federal government, through the efforts of the U.S. Forest Service, is still best suited to managing and conserving lands like the Nantahala National Forest.

James G. Lewis, PhD
Forest History Society, Durham, North Carolina

LETTER FROM A DISTRICT RANGER

Greetings from a retired U.S. Forest Service employee—but, more importantly to me, a former district ranger on the Nantahala National Forest. Soon after arriving in Franklin in 1980, I met retired Forest Service researcher Walton Smith. His early words let me know that I was now working on the best national forest in the country. In my first week, employee Rebecca McConnell asked me to go outside to see something. She pointed out Standing Indian Mountain and Albert Mountain Fire Tower. What a beautiful sight on a clear, cold day! But what really impressed me was how these folks felt about "their" forest.

The diversity of the Nantahala is amazing. High mountain peaks, deep clear lakes, whitewater rapids, spectacular waterfalls, miles and miles of trails, wilderness areas and many developed recreation sites are only a portion of the diversity. But equally important are the Beech Creek Seed Orchard, the world-famous Coweeta Hydrologic Lab, the Lyndon B. Johnson Job Corps Center, majestic fire towers, the Joyce Kilmer Memorial Forest and many historical sites.

Who maintains all the trails? Who manages the Job Corps Center and is involved in helping young people become productive adults? Who fights the wildland fires? The questions could go on and on, but the point is made. Dedicated employees and many partnering volunteers are the lifeblood of the Nantahala. Volunteers are the primary trail maintainers on the hundreds of miles of Nantahala trails, saving thousands of dollars for the Forest Service. This includes foot, horse, bike and off-road vehicle trails. Volunteers also serve in offices, in developed recreation sites and at the LBJ Job Corps Center.

Forestry technicians are generally the employees in non–college degree positions. In my opinion, these are the most important Forest Service employees, along with the office staff. Never tell or hint that there is a job that can't be done. Just get out of their way! A few times in my career on the Nantahala, I simply hinted at projects that *should* be done. A suggestion was made to construct a roof on Wayah Bald Tower. I went by there nine weeks later, and it was 90 percent completed. A clothes-changing room design was described on the back of an envelope for the Nantahala River Launch site. I went by there six weeks later, and it was built! Good teamwork is essential, and the forestry technician and office staff made my ranger district function at a very professional level…and made this ranger look really good.

LEWIS KEARNEY
Retired Wayah district ranger, Nantahala National Forest

ACKNOWLEDGEMENTS

Sounds like you can't get your wagon unloaded, lady!" said Johnny Shields, president of the Smokey Mountain Off-Road Vehicle Club.

Nantahala's interviews and stories swirled around inside my head while I searched for the perfect niche for each of them in the written preservation of Nantahala's history. It was time to pull it all out of my head, off my notes and out of my resources and typed into the final draft of the manuscript. The deadline for the book drew near, while return communications, reviews, edits and follow-up interviews mounded up. The main reason I was having such a difficult time getting my large Nantahala research wagon unloaded, though, was because it was so overloaded by the tremendous support, inspiration and assistance from a forest-full of people who love this national forest and its history.

The history of Nantahala National Forest is now typed on pages and bound in a book, but its stories belong to those who have lived them, to those whose ancestors spoke them and to those whose spirit connects to their deeper meanings. After listening and learning, reading and researching, exploring and getting lost, I have compiled the individual stories into one cohesive, historical unit. My hope is that my written words have retained the special personalities that shared them—personalities like retired USFS forester Marshall McClung. On the table in front of me at a diner in Franklin, he thumped down two volumes of his books that he had written about Graham County history.

"Here, these might help," he said quietly. I felt like I had been given a pot of gold, but the gesture may have been even more meaningful than his valuable historical information. Offering the personal lifetime accumulation of his articles written about Graham County was the emotional jump-start and support I needed to dive into the history of another national forest.

Retired USFS employee Jean Saunders had invited me to lunch in Franklin with Forest Service retirees to discuss my recently released book on the history of Pisgah National Forest and my plans to write one about Nantahala. Former USFS employees in Nantahala meet for lunch each month, like family. "You make new friends, but you never lose friends in the Forest Service," said Scooter Brown, an eighty-year-old retired USFS district ranger. Thank you, Jean. I owe you and several others a great deal of gratitude for initiating a chain-reaction response of resources and contacts eager to share Nantahala's history.

Before beginning his on-call North Carolina State Forest firefighting shift one weekend, Marshall McClung met me at 6:30 a.m. for a guided hike to the Denton homeplace in Joyce Kilmer Forest on Saturday and met me for a two-hour interview on Sunday. Throughout the one-and-a-half-year research process, he answered e-mail questions without delay. Former USFS interpretive specialist Bill Lea, now professional wildlife photographer, e-mailed massive amounts of interviews and historical files he had accumulated during his service with Nantahala National Forest. Retired district ranger Lewis Kearney must have cleaned out his attic to flood my mailbox with newspaper articles, documents and other historical resources; sat for several lengthy interviews; and patiently answered my endless e-mailed questions. Retired district ranger Joe Bonnette kept me updated on Cheoah District activities and invited me to meetings, hikes and other events, where I met Kim Hainge, Jim Kriner, Joey Muehlausen, Lisa Russo, Dick Evans, retired USFS district ranger Scooter Brown and others, who provided valuable information. Retired USFS resource assistant Chad Boniface dedicated three full days to touring me around the Nantahala Ranger District, as if he had all the time in the world. Retired USFS assistant ranger Bruce Bayle joined us on one tour to provide his insight. Chad also arranged for a guided off-road vehicle tour of Wayehutta with Ronnie Dietz, former landscaper/groundskeeper at Western Carolina University and now the volunteer gatekeeper-greeter for Nantahala's all-terrain vehicle course. The president of the off-road club, Johnny Shields, met me at the Moose Café for an interview. Retired USFS forest technician Hoot Gibbs offered his experience and knowledge via e-mails and retiree luncheons and contributed his article of the "must see" places in Cheoah. Thanks, also, to USFS fire management officer Chad Cook for taking me on a tour of the seed orchard in the Tusquitee District; to retired USFS seed orchard manager Glen Beaver for the fabulously fun interview over gravy biscuits at Bojangles; and to retired USFS employee Clay Logan for an entertaining meeting in the 'possum capital of the world. I sincerely appreciate the valuable education you provided, Robin Taylor, current USFS seed orchard manager, on the genetic resource area.

Thanks to my daughter and granddaughter, Christi Worsham and Brooke Lyda, for exploring Nantahala's trails with me, for Easter and Thanksgiving at Fontana Village and for Mother's Day at Highlands. I could not have stayed focused and dedicated to the task without the continued support and encouragement from my husband, John Spencer. He calmly endured all of our crazy drives in the backcountry, our aching necks from long hours bent over documents at libraries and museums and my weeklong writing sessions spent from sunup to sundown in my library upstairs.

Considerable assistance came from curators, docents, historians, authors, librarians, researchers and others who seemed as obsessed with Nantahala's history as I am. Robert Shook rolled out the red carpet at the Macon County Historical Museum. Wanda Stalcup did the same at the Cherokee County Museum, spending hours searching for historic photographs. The staff at the Hayesville County Library gave me undivided attention and assistance. The Highlands Biological Station opened the research library for me to peruse at leisure. In Highlands, the fine ladies at Hudson Library spent extra time helping me locate resources. Pat Griffin, the knowledgeable docent at the Highlands Preservation Society, was so informative and charismatic that it was hard to leave. Kathy Flowers at the USFS Coweeta Hydrologic Laboratory provided a grand tour and an armful of documents and research studies. The phone call from the retired Coweeta scientist Dr. Wayne Swank provided much-needed support and information.

I also owe many thanks to the staff at the libraries at Sylva, Western Carolina University, University of North Carolina–Asheville, Andrews, Robbinsville, Franklin, Cashiers, Asheville, Murphy and Andrews. I also want to thank the staff at the Graham County Visitor Center, the Museum of the Cherokee Indian, the Cherohala Skyway Visitor Center, Fontana Village, Glenmary Missioners Archives, the Junaluska Memorial and Museum, the Forest History Society, TVA's Fontana Dam and Visitor Center and the Trail of Tears National Historic Trail Museum in Murphy.

My words have been polished, corrected, clarified, critiqued, refined, edited and improved by dozens of people who reviewed sections of the manuscript related to their individual interests and expertise. I wholeheartedly thank Mike Ingram, Gretchen Kirkland, Carolyn and Buster Stewart, Leota Denton Wilcox, Kim Hainge, Jenny Wilker, Donna Duffy, John Cottingham, Lewis Kearney, Bob Scott, Chad Boniface, Johnny Shields, Marshall McClung, Hoot Gibbs, Dick Evans, Robert Rankin, Gary Kauffman, Dan Pittillo, Ila Hatter, George Ellison, Gerald Ledford, Robin Taylor, Glen Beaver, Clay Logan, Chad Cook, Michael Gora, Ron Taylor, John Lane, Bill Champion, Terry Pierce, Wally Avert, Joanna Padgett-Atkisson, Marilyn Reid, Burt Kornegay, Brent Martin, Barbara Duncan, Lamar Marshall, Lisa Whaley, Bill Nichols, Clint Horton, Karl Sutter, Lisha and Lindy Ammons, John R. Ray, Payson Kennedy, Will Leverette, Lloyd Swift, Wayne Swank, Kathy Tinsley, Laurel Radley, Callie D. Moore, Albert Rufus Morgan III, William Van Horn, Gail Lehman, Bill Lea, Christine Kelly, Randolph Shaffner, Lucy Putnam and Representative John Ager.

I extend heartfelt gratitude to my good friend and "coach," retired Clemson University professor Jere Brittain, who read every word of the manuscript during the creative process, even the roughest and most tedious of early drafts. Jere helped me restructure sentences, rethink ideas and consider different points of view. Just as valuable has been the time and dedicated interest devoted to this project by James Lewis, PhD, with the Forest History Society. He, too, read the entire text, providing generous constructive critiques, professional advice and knowledgeable insight into the history of forestry. A published history may be enlightening and entertaining for the reader, but with the help and guidance of others, the writing process becomes an in-depth educational and personal enrichment experience for the author as well. Thank you, everyone.

I once again thank my commissioning editor, J. Banks Smither at The History Press, for believing in me and Nantahala National Forest like you did with Pisgah National Forest and Clingmans Dome. Thanks to the image design team for capturing the beauty of Nantahala and its history in the photographic layout. Thanks to my copyeditor at The History Press, Ryan Finn, for scrutinizing the text and providing suggestions for improvement.

I want to offer a special expression of gratitude to Barbara Duncan, PhD, education director of the Museum of Cherokee History. A deep Cherokee heritage and presence exist throughout the Nantahala region. Representing that rich, profound culture with respect, dignity and accuracy was of paramount importance to me. Thank you, Ms. Duncan, not only for reviewing and improving my words regarding Cherokee history but also for sharing your beautiful poem, insight and knowledge.

PART I
THE STORY OF NANTAHALA

We heard a noise like an airplane flying real low," reported a fisherman to the sheriff of Macon County, "and then it went silent."

"Then we heard a loud bang," his buddy added.

"When did you hear this?" the sheriff asked.

"Yesterday."

"Yesterday?"

"Yeah, we were fishing at Nantahala Lake. We fished all day and didn't get back home until after dark."

Shortly after noon on September 12, 1985, Sheriff George Moses called District Ranger Lewis Kearney. "I need help with a search party," the sheriff told him. "Sounds like a plane went down yesterday in the national forest west of Nantahala Lake."

U.S. Forest Service (USFS) personnel and county deputies gathered near Big Choga. Sheriff Moses, Ranger Kearney and a search party hiked west up a steep trail toward the ridge forming the boundary between Macon and Clay Counties. Radio communication between Kearny, a USFS dispatcher and the Federal Aviation Administration (FAA) confirmed the presumption of a downed plane in Nantahala National Forest (NNF). "If you find it, call us back with the plane's tail number," said the FAA official.

Meanwhile, Clay County sheriff Tony Woody also received the report of a plane crash and organized two search parties. *Asheville-Citizen Times* reporter Bob Scott joined the group that got lost. "To this day, I still don't know where we were," said Scott, now mayor of Franklin.

Sheriff Woody and a private aircraft pilot departed a small Tusquitee airstrip for an aerial view. Near Weatherman Bald (4,960 feet) and County Corners (5,419 feet), where Macon, Clay and Cherokee Counties meet, they spotted the wreckage. The second Hayesville rescue group arrived at the crash site and discovered scattered airplane debris around an intact instrument panel. They searched the scene for survivors. Somewhere in the wilderness, Scott heard their radio transmissions.

"How many people were killed?" the airborne sheriff asked the ground crew.

"Sheriff, we can't find any bodies."

"Now, boys. This is serious business. That plane didn't get there by itself."

The Hayesville Rescue Squad was combing the scene when the Macon County search party arrived. Sheriff Moses and his team helped search the wreckage for the pilot's body. There were no signs of a fire. No smoke. No explosions. And no body.

Kearney radioed the crash scene report of the evidence found—and not found—to the FAA. The FAA contacted the Federal Bureau of Investigation (FBI). The search crews departed the rim of Fires Creek and hiked down into Clay County after dark, presumably descending into the watershed on the Shinbone Ridge trail to the end of FSR 340. The next day, federal agents and other officials arrived to investigate the scene.

Mystery and intrigue; conspiracy and international crimes; a compelling plot with an unusual twist or two; drugs, weapons and a plane crash with a missing pilot—it all sounds more like a dramatic mystery made for a movie than the beginning of a nonfiction book on the history of a national forest.

At 8:44 a.m. the previous day, a Knoxville resident found a dead man, dressed in combat-style fatigues and expensive Italian shoes, lying in his front yard next to a partially opened parachute. In a green army duffel bag, the man had $15 million worth of cocaine. Authorities also found knives, night-vision goggles, rope, automatic weapons, nearly $5,000 in cash, six South African gold coins, a personal address book and keys to an airplane. The plane's identification number on his keys matched that of the twin-engine Cessna-404 found in NNF.

A Knoxville air traffic controller had tracked an unidentified airplane on radar on September 11. The pilot, however, had not filed a flight plan or made radio contact with the tower. The radar had traced its flight path looping around Knoxville and heading back toward North Carolina on a southeasterly heading. Investigators learned later that the forty-year-old pilot, Andrew Thornton II, a former Lexington, Kentucky narcotics officer and an attorney turned drug dealer, parachuted to his death before his Cessna, placed on autopilot, ran out of gas. While in the army in the 1960s, Thornton was an experienced paratrooper with the 101st Airborne Division. The government awarded him a Purple Heart for injuries received during the Dominican Republic rebellion. Later, in the 1970s, after a six-month prison sentence for marijuana trafficking, Thornton became a leader of "The Company," a Kentucky-based drug smuggling ring. More than three hundred members, including policemen and politicians, owned more than $26 million in boats and planes. That September in 1985, his plane, equipped with long-range fuel tanks, was carrying him back from Colombia in South America.

Two days after the search parties located the Cessna in NNF seventy miles south of Knoxville, USFS agents found 220 pounds of cocaine ($56 million worth) about fifteen miles farther south in Chattahoochee National Forest. Beside a large meadow, army duffel bags dangled from a parachute stuck in a tree. The locked duffel bags held zippered, black nylon bags. Those contained individually wrapped packages of cocaine marked "USA 30" or "USA 10." The packaging and markings matched those found on the pilot.

On December 23, 1985, the *New York Times* reported that "a 175-pound bear apparently died of an overdose of cocaine after discovering a batch of it." Surrounding the black bear in a forest outside Blue Ridge, Georgia, forty opened packages with traces of cocaine also matched those carried by Thornton.

The stories of search-and-rescue operations in the mountains of Western North Carolina could fill up the pages of their own book about the history of NNF. Inclement weather, slippery waterfalls, disorientation and inexperience have caused serious and sometimes fatal mishaps. However, the Cherokee Nation embraced this rugged area of five-thousand-foot peaks and dense forests as their home. The remote Southern Appalachians also attracted white settlers, explorers and an agency forming a new national forest.

The Appalachian Mountains stretch from Maine to Alabama. From the north, the Southern Appalachians enter North Carolina and divide into parallel ranges. The Unaka Mountains border Tennessee. The Blue Ridge Mountains lie seventy miles east of the Unakas. In the southern Unakas, known as the Unicoi Mountains, several cross-ranges—like the Balsams, the Cowees, the Nantahalas and the Snowbirds—span the area between the parallel ranges as if reaching toward its twin range in a geological gesture of friendship.

In a 1953 edition of *The State*, Bill Sharpe compared the Southern Appalachians to a ladder:

> [They] *arrive at the North Carolina border, after a prim and proper journey through Virginia* [and] *things begin to happen...makes geographers and geologists splutter with explanations and reach for unscientific adjectives...a maze of valley, ridge and river bemuse the traveler and send him home hardly knowing where he has been....Try some unscientific thinking. Here is a ladder, with the eastern post representing the Blue Ridge, the western post the Unakas. Connecting these twin ranges of the Appalachians are tremendous cross-ranges, bridges of land,* [like]...*the rungs or steps of the ladder. The rungs separate river basins, each with its own elaborate water system. The rungs do not always go across directly and completely. Some are bent; and there are splinters, both on the upright posts and the rungs. And even the splinters have smaller splinters.*

In the southwestern corner of North Carolina, the Nantahala region is one of the wettest regions in the country—second only to the Pacific Northwest. As headwater streams gather volume and force, they carve narrow valleys, deep ravines and winding channels. Streams join rivers, like the Tuckasegee, Valley, Whitewater and Cullasaja. The Nantahala River changed its course over time, carving a gorge through softer limestone. The Chattooga River, in part, flows freely as a national wild and scenic treasure. The basin of the Little Tennessee River attracts scientists for watershed research. Power companies have dammed other rivers—like the Hiwassee, Cheoah and the Little Tennessee—creating scenic mountain lakes. Some of Nantahala's streams tumble over ledges into gorgeous cascades and waterfalls, such as 411-foot Whitewater Falls, the highest in the East. West of Pisgah National Forest and south of Great Smoky Mountains National Park, these 500,000 well-watered acres make up Nantahala National Forest.

With its vast aquatic resources, Nantahala's rivers influenced the location of settlements and villages established in Western North Carolina, like the Cherokee settlements of Nikwasi and Cowee on the banks of the Little Tennessee. In the early 1700s, South Carolina colonists established trade with the Cherokees in the Nantahala region, exporting deer hides traded by the Cherokees from Charleston. From 1721 to 1777, treaties with the Cherokees ceded lands in South Carolina, Georgia, Virginia and North Carolina, including parts of their Nantahala homeland. The Treaty of Hopewell, signed in 1785 with the new American government, ceded

Big Santeetlah Creek, Cheoah Ranger District. *National Forests of North Carolina Historic Photographs, D.H. Ramsey Library, Special Collections, University of North Carolina–Asheville.*

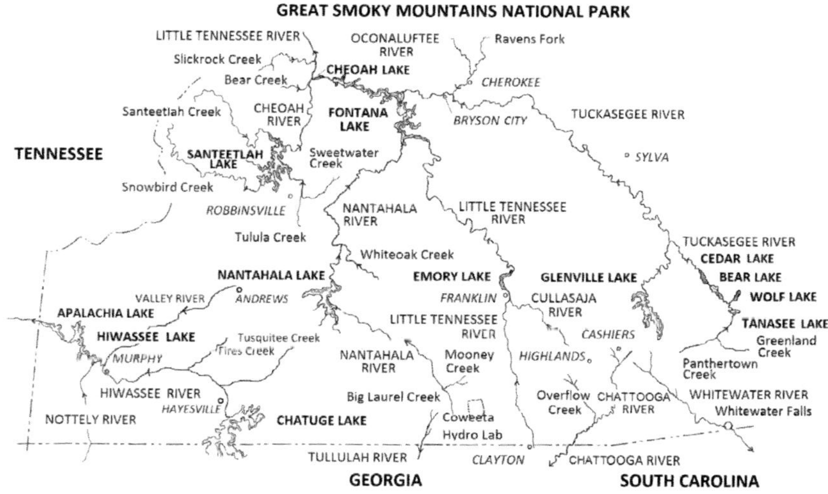

Watersheds of Nantahala National Forest. *Illustration, Ken Czarnomski, architect/illustrator/cartographer/naturalist, phoenix2reach@gmail.com.*

more land in North Carolina, as well as tracts in Kentucky and Tennessee. Negotiations, treaties and battles between the Cherokees and the federal and state governments continued for the next three decades.

South Carolina opened former Cherokee territories to white settlers, offering some tracts as land grants to Revolutionary War veterans. In 1787, the Chattooga River became the boundary between South Carolina and Georgia. By 1807, commissioners wanted to mark the line between Georgia and North Carolina. In a meeting at the Buncombe County Courthouse in Asheville, North Carolina, officials agreed to start the boundary line at the juncture of North and South Carolina at the Chattooga River and run due west along the 35th meridian. Georgia's governor hired Major Andrew Ellicott in 1811 to survey the rough terrain. On the east bank of the Chattooga River, Ellicott inscribed a rock with the initials "N-G," for "North Carolina–Georgia." In an 1813 dispute, commissioners from North and South Carolina hired other surveyors to reassess the area. The surveyors chiseled a new cornerstone, several feet downstream, with the inscription "LAT 35 AD 1813 NC+SC." Known as "Commissioners Rock" and recognized today as the true tristate juncture, hikers often misidentify it as "Ellicott Rock." In 1973, the district ranger of Nantahala National

Forest completed documentation to have the site designated as a national historic site.

Disputes between landowners along the state line continued, especially in the early 1900s, when the U.S. government expressed interest in buying unwanted land for a new national forest. Some claimed that locals had moved the state survey markers; others believed that landowners had moved Commissioners Rock. A 1926 USFS map describing the creation of NNF notes, "The Forest Service location of the line established the fact that Commissioners Rock had not been moved, but also determined that the line does not follow the 35th degree, being some distance south of this latitude."

A USFS employee inspects Commissioners Rock, marking the intersecting point between North Carolina, South Carolina and Georgia. *USFS.*

In 1819, another federal treaty ceded a large amount of the Cherokees' remaining land, leaving them with small remnants of their homeland in parts of Georgia, Tennessee and Western North Carolina. People of European descent moved in quickly. By 1820, settlers occupied vacated lands in what is now Macon County. New settlers bought property at public auctions in Waynesville. They built homes, barns and roads. They cleared land for agricultural fields and harvested timber for wagons, firewood, building supplies and other items. Between 1829 and 1830, Franklin built its first courthouse. In 1830, President Andrew Jackson's policy to remove the Cherokees from the Southern Appalachians passed in Congress by a slim margin. In 1838, the Removal began. At least four thousand of the fifteen thousand Cherokee men, women and children forced to march to Oklahoma on the Trail of Tears died during the journey.

By 1891, Southern Railway's Murphy Branch rail line had opened the region to miners, businessmen, entrepreneurs, lodge owners and visitors. The State Board of Agriculture and tourism departments promoted the mountains' scenic beauty and health benefits. The hardwood forests found in the Southern Appalachians attracted northern logging companies. Loggers built sawmills near railroads to ship the lumber to markets. Portable mills processed lumber at one site and then moved to greener ones. Mountainsides became bare and eroded. Discarded brush caught fire and inflamed acres. Anxious citizens and early conservationists began voicing concern.

When Dr. Chase P. Ambler moved from Cincinnati to Asheville in the late 1800s, he fell in love with the Western North Carolina mountains. Disturbed by the vast destruction of the region's forests, he vowed to protect them. He advanced the idea that some of the forests should belong to the people as a national forest or a national park and not be destroyed and exploited for the monetary gains of a few.

In 1899, Ambler helped create the Appalachian National Park Association. At a meeting that November, North Carolina governor Locke Craig declared, "The Government must preserve this valuable gift of nature for the benefits of all the people and now is the accepted time." In a letter to Congress pleading for the government's help, leaders of the association wrote, "The Southern Appalachians are of unsurpassed attractiveness…the national government… by methods of scientific forestry [can] preserve the forest as a heritage and a blessing to unborn generations….It is the duty of the national government to protect the water supply of the country."

In 1901, after Congress approved $5,000 to fund a survey of the southern forests, H.B. Ayers of the U.S. Geological Survey and W.W. Ashe of the U.S. Division of Forestry evaluated the geology, topography and hydrology of the forests in the Southern Appalachians. Their studies included the region's climate and the conditions of its flora and fauna. Ayers and Ashe found widespread damage from heavily logged areas and poor forestry practices. Their reports described extensive areas cleared of trees for agricultural needs. They recorded eroded hillside farms robbed of valuable topsoil. Barren areas caused silt-laden rivers to flood. About 80 percent of the surveyed forests had suffered fire damage. In some cases, landowners had ignited fires to control weeds and insects, harming young trees and scorching fertile soils. The secretary of agriculture, James Wilson, issued a final report based on the findings of Ayers and Ashe and the conservation concerns of American citizens. In his report to President Theodore Roosevelt, Wilson wrote, "Interest in practical forestry has grown up among the American people…our country shall profit by the preservation of our forest remnants before it is altogether too late."

A decade earlier, Congress had passed the 1891 Forest Reserve Act. This bill allowed the president of the United States to establish forest reserves carved out of the public domain. President Benjamin Harrison created the first forest reserve in Wyoming and then others in Colorado, California, Oregon and the territories of Alaska, Arizona and New Mexico. By 1907, the new U.S. Forest Service had changed the name of forest reserves to national forests. Out of lands in the public domain, the government formed the Arkansas, Ozark, Ocala and Choctawhatchee National Forests in the Southeast. However, very little public land remained in other southeastern states. The secretary of agriculture urged "the government to protect these Appalachian forests by purchasing the lands and making of them a great national forest....The States of the Southern Appalachian region own little or no land, and their revenues are inadequate to carry out this plan. Federal action is obviously necessary."

Repeated proposals, granting the government the power to purchase lands to form national forests, were introduced into Congress but failed to pass. When the Appalachian National Park Association learned that Congress preferred a national forest, where forest management would help meet the timber needs of a growing population, rather than a national park, where policies prohibited timber harvesting, Ambler's group changed its name to the Appalachian National Forest Reserve Association. However, after thirty years and a passionate public campaign, Congress would not budge. By 1907, the American Forestry Association had decided to give a national voice to the cause. Gifford Pinchot, an ardent conservationist and chief of the new USFS, also supported the movement to protect eastern forests. The citizens of New Hampshire, with the support of Massachusetts congressman John W. Weeks, who summered in their mountains, had also campaigned for similar protection for lands in the White Mountains. When a second report on the southern forests was published in 1907, it confirmed the disturbing results of 1901. Armed with those results, the Forest Service began identifying regions of interest.

After a decade of debate, Congress passed the Weeks Act, which authorized land purchases that would protect the headwaters of navigable streams. In 1911, the act protected the forests' natural resources and provided for management of forest fires. It also established the National Forest Reservation Commission to select appropriate tracts for the new national forests. State governments passed enabling laws, permitting the federal government to buy land in each state. The USFS had finally secured the approval and appropriations needed to begin acquiring tracts of land that it had identified for protection.

When President William Taft signed the Weeks Act on March 1, the USFS was ready. Forest agents had already identified potential purchase units, parcels of land in need of management and rehabilitation. Large purchase units often contained many thousands of acres. Some tracts held high-quality timber in virgin forests, while others were barren forests left by loggers. By March 27, the National Forest Reservation Commission had approved the first purchase units to become national forests in the East. Many were in North Carolina. By 1916, the USFS had joined two purchase units to form Pisgah National Forest, the first national forest in the East created under the Weeks Act. By 1918, the USFS had created Alabama National Forest (now Bankhead National Forest) in Alabama, Natural Bridge and Shenandoah National Forests (now George Washington National Forest) in Virginia and the White Mountains National Forest in New Hampshire.

On January 29, 1920, President Woodrow Wilson combined purchase units with lands in Georgia and in North and South Carolina to form Nantahala National Forest. "The inclusion of the Nantahala region [as part of the national forest system] followed naturally," a 1936 USFS booklet stated. "Nowhere was there greater or more urgent need for governmental protection of the mountain watersheds against wanton timber waste, burning, and soil erosion. The work of acquiring and consolidating the Nantahala National Forest commenced promptly after the passage of the law and appropriations of funds and has gone forward steadily."

In creating NNF, President Wilson had excluded USFS lands in Swain, Graham and Cherokee Counties, North Carolina, so President Harding placed these lands under the management of Pisgah National Forest the following year. Former USFS district ranger William L. Nothstein explained, "For a time, custodial responsibility was placed with the Pisgah National Forest Supervisor in Asheville rather than the Nantahala National Forest Supervisor in Franklin. Even though Franklin was much closer, travel to and from Asheville by railroad was more dependable than Model-T Ford transportation from Franklin over poor wagon roads." By 1929, President Hoover had transferred these lands in the western counties to Nantahala National Forest, expanding its boundaries.

In 1925, the National Forest Commission relinquished a key tract of land on the North Carolina–Tennessee border that it had listed as a potential national forest. Preservationists were proposing a new national park in that region. Releasing the Smoky Mountains Purchase Unit from USFS consideration allowed the National Park Service access to one of the last remaining contiguous, majestic forests in the East. By 1926, President Coolidge had approved the creation of the new Great Smoky Mountains National Park, which included the former USFS purchase unit.

By the time President Taft had signed the Weeks Act, logging companies in the Nantahala region were ready to sell. Within one month, Gennett

Lumber Company of Atlanta made an offer to sell 31,000 acres across four Georgia counties to the government. Within a few years, the company sold 20,000 more acres. The Macon County Land Company sold 16,000 acres to the Forest Service. Between 1912 and 1915, Macon Lumber Company sold the agency 21,000 acres. The government's average price was $6.91 per acre (a range of $2.00 to $12.00 per acre). Since the Macon Lumber Company had only logged 102 of its inaccessible, mountainous acres, it received $11.00 per acre.

Lumber companies continued selling large tracts of land throughout the 1920s and the 1930s. After harvesting marketable timber, lumbermen wanted to sell out before property tax deadlines. A severe chestnut tree blight sent others packing. So did the Great Depression. With an eager buyer like the USFS wanting to buy huge tracts of unwanted property, selling out was both easy and profitable to the lumber companies. Private owners also sold acreage to the government. In one region, the USFS purchased 850 tracts of land from 359 private owners. Not all owners wanted to sell, however, leaving private inholdings scattered within large tracts of USFS land.

Major remapping of the USFS landscape occurred in 1936. Nantahala's South Carolina tracts became the new Sumter National Forest. Chattahoochee National Forest took over the lands in Georgia, and Cherokee National Forest managed the forests in Tennessee. On the Tennessee–North Carolina line, the USFS eliminated Unaka National Forest and divided its acreage between the national forests of Pisgah, Jefferson and Cherokee. In the 1940s, after the Tennessee Valley Authority (TVA) completed its dams for the Chatuge, Hiwassee and Apalachia Lakes, it transferred thousands of acres to the USFS. By the 1950s, agencies had again redefined the boundaries of the Great Smoky Mountains National Park, as well as Nantahala, Pisgah and Cherokee National Forests.

Throughout the decades, the USFS has reshuffled its parcels of land in Western North Carolina like unmatched pieces of a puzzle. Numerous surveys, exchange agreements, donations, additions and deletions of land units have redrawn the map of NNF. To manage its resources efficiently, the Forest Service frequently attempts to consolidate its holdings, explained Chad Boniface, retired forester for the former NNF Highlands Ranger District. A high priority is to obtain "key recreational and scenic areas," as well as "wildlife inholdings" that make regions more accessible to the public and reduce habitat fragmentation in the forest's ecosystems. Available lands lying adjacent to or within current USFS holdings, said Boniface, garner the greatest interest. In a 2015 public meeting for the revision of the Pisgah/ Nantahala National Forests Comprehensive Management Plan, USFS staff reported that "about one-half of the Forest Service lands acquired since 1992 are directly adjacent to currently designated boundaries."

Since the passage of the Weeks Act in 1911, part of the Forest Service's mission in the East has been to purchase lands to protect important watersheds and manage the forest's natural resources. Following the passage of the Multiple-Use Sustained-Yield Act of 1960 and the National Forest Management Act of 1976, the USFS now balances its management of watersheds, wildlife, range, minerals, timber and recreational use. The agency practices improved logging methods and research-based forestry principles to provide a sustainable yield of the public's need for wood products for generations to come. To prevent erosion, it plants trees and reseeds road banks. It surveys and maintains boundary lines, builds roads and restores recreation areas. To benefit wildlife, the USFS monitors and maintains valuable ecosystems.

"What makes most former NNF employees the proudest," said retired USFS district ranger Joe Bonnette, "is the recovery effort and the reforestation of the 'lands that nobody wanted.' Our forests in the south were severely overcut, abused and eroding away when the USFS obtained them and it took decades of recovery work and management for them to be as beautiful, relatively healthy and productive as we see them today."

Since its early days, the USFS has emphasized the importance of forest fire control, management and education to protect the natural resources, as well as the public and its personal property. Fire safety education for the next generation is a USFS top priority. At Junior Ranger Camps, USFS Marshall McClung and Joe Bonnette illustrate how fires spread through a forest by lighting matches, poked into a pegboard like trees rooted on a hillside. At times, wildfire education comes in a big, furry package named Smokey Bear.

"It must have been 140 degrees in there [inside the suit]," said McClung. "It was the Fourth of July parade; I was drenched with sweat and couldn't wait to take my head off." As Smokey Bear jumped off the USFS truck at the end of the parade, "forty-nine adults and kids came running at me!"

"Smokey!" they said. "I want a photo with Smokey!"

"I just turned around and hugged them all," said McClung. "People love Smokey. I was baking inside that bear suit, but Smokey was smiling from the inside out!"

From October to December 2016, wildfires burned more than 118,000 acres in eight states in the Southeast. Some areas had received less than

USFS Joe Bonnette and Marshall McClung demonstrate wildfire behavior at a Junior Ranger Camp at Tri-County Community College. *Kim Hainge, godsgardens@earthlink.net.*

two inches of rain in three months; others recorded no rain. The relative humidity remained low. Dead forest materials routinely exchange moisture with the air. When the humidity is high, forest fuels siphon moisture from the atmosphere. When the humidity is low, the air takes water from the fuels. Pine needles, fallen leaves and other light fuels lose moisture quickly to dry air, creating a tinderbox of highly flammable fuels. Heavy fuels, like trees and shrubs, dry out more slowly. In the Southeast, more than six thousand firefighters from Puerto Rico to Alaska and across the continental United States battled the blazes, sparked by human activity or lightning bolts and occasionally intensified by strong winds.

In NNF, wildfires scorched nearly twenty-five thousand acres, including lands near Calderwood and Santeetlah Lakes and the Cherohala Skyway; in the Standing Indian, Wayah Bald, Rufus Morgan Falls, Chunky Gal Mountain, Joyce Kilmer–Slickrock and Southern Nantahala Wilderness areas; and in the Whitewater Falls, Buck Creek, Bartram Trail, Nantahala Gorge, Tellico and other regions. During the active wildfire period, fire management teams analyzed each region's topography, fuel load and weather conditions, as well as the fire's progression and the effectiveness of suppression efforts. Dozer lines protected structures and personal property.

Road and trail closures and evacuations helped ensure the public's safety. Crews maintained and patrolled fire lines, cleared containment lines of freshly fallen leaves and ignited back-burning fires to limit the speed and spread of the wildfires. They addressed flare-ups and hot spots. When weather, visibility and resources permitted, helicopters dipped 2,800-gallon "Bambi buckets" into Chatuge, Fontana and other Nantahala reservoirs and drained its earth-quenching relief through eighteen-inch holes in eight seconds at the fire's edge. Chinook helicopters, Helitank Sky Cranes and other aviation resources doused interior flames to reduce the fire's intensity. Nighttime infrared aerial cameras searched for heat amid the cooled, blackened forest. Hand crews focused on removing flammable fuels at significant sites, such as the giant poplars on the upper loop of the Joyce Kilmer Memorial Trail; raking leaves; removing fallen trees; and dynamiting dead hemlocks. For the safety of the firefighters, operation managers evaluated an area's steep terrain, muddy roads and hazardous, falling trees.

On November 14, the USFS and North Carolina State Forest Service opened the Joint Information Center (JIC) as a "one-stop center to provide the public with updates on wildfires, evacuations and shelters, road and trail closures, air quality, current burning restrictions and the schedule for public meetings." Communities offered generous food, water and moral support for the firefighters. In Hayesville, five hundred schoolchildren waved banners and American flags as firefighters departed for work from their schoolyard. The children yelled, "Thank you!" and the firefighters saluted back.

By December 6, the USFS had reported that all fires in WNC were contained, except two. Nature's steady rains helped with those. Crews continued mop-up procedures and securing containment lines while the post-fire assessment and restoration work began. Since extensive wildfires and fire-suppression operations can lead to floods, erosion and sedimentation in streams, the USFS in North Carolina established a Burned Area Emergency Response (BAER) team. BAER would evaluate the needs of the forest and identify short- and long-term restoration goals to address public safety and protect cultural and natural resources.

At the time of this writing, trail clubs had begun helping the USFS assess the damage to shelters, trail-treads and water bars. Civic leaders began campaigning for funds to rebuild the wooden cab on Wayah Bald Fire Tower, the wooden staircase on the Foothills Trail at Whitewater Falls, the observation deck at Maple Springs Overlook and other projects. Volunteers had hiked Kilmer's Memorial Trail, finding damaged handrails, severe erosion and dangerous standing trees weakened by the fire. Humans will offer time,

NAMING PLACE

By Barbara Duncan, PhD, executive director of the Museum of the Cherokee Indian; first published by Voices of the American Land

English place names
clatter on our tongues
cacophonous gibberish:
Soco
Oconaluftee
Tusquittee.
They mean:
nothing.
They simplify:
people were here, now gone.
The names remain, shadows.
Cherokee place names
speak truth musically:
sogwo ageyv'i
one creek
carves miles of valley;
egwoni nuhladi
fast river
dashes downhill from Kuwahi's creeks and springs
fifty miles to flat river bottom;
tusquittee—the house rafters in the old days,
shape of mountains above Hiwassee.
People—still here.
Mountains—still here.
Words—still here.
If you want to know the truth:
unpave the roads and walk old trails;
root up parking lots and see villages;
live here a thousand years
until every breath sings the mountains.

labor and funds to restore the forests, but throughout the scorched acres, nature, too, will exert its curative powers. By May, the fresh green forest floor will flourish with an array of wildflowers.

The story of Nantahala National Forest is not just a tale of forest fire operations, land acquisitions, search-and-rescue efforts, Cherokee heritage, white settlers, logging and USFS management. It includes the story of schoolchildren, a school nurse and a cycling club creating one of the best biking courses in the region. Nantahala's story includes conservationists delaying the dreams of a wagon train to connect towns in Tennessee and North Carolina with a scenic skyway; stopping a Blue Ridge Parkway extension from destroying a beautiful valley enhanced by granite domes; protecting the waters of a wild, free-flowing river; and campaigning for protection of mountain camellias. It's about erecting telephone poles that help endangered flying squirrels glide safely over a highway. It includes archaeologists and historians searching for ancient Cherokee weirs and footpaths and identifying the sites of forts and camps that held Cherokees forced from their homelands.

Nantahala's history also includes scientists and others who have followed the footsteps of William Bartram and other early explorers to identify flora and fauna. In the Nantahala area exist some of the rarest species in the world, like the noonday globe snail and the Nantahala spider. Botanical societies, universities, hiking clubs and volunteer groups lead guided walks in the region's wild gardens. Families, hikers and backcountry explorers seek new discoveries on Nantahala's trails, while others maintain them. Students and professors study Nantahala's trees, topography and geological formations. Rockhounds search Nantahala's Cowee and Chunky Gal Mountains and the serpentine pine barrens for olivine, corundum and other minerals.

The stories of mountain lodges operated by Catholic priests at Buck Creek, innkeepers in the Snowbird Mountains and power company officials on the Cheoah River have become part of Nantahala's history. Its past also holds humorous tales, like army soldiers misbehaving at Appletree Campground, prompting the base commander's surprise visit. Nantahala's recorded history preserves other personalities, like a reverend who single-handedly maintained sixty miles of the Appalachian Trail (A.T.) for decades, a librarian who changed careers and helped create the largest whitewater training center in the United States and a real estate developer who saved Whiteside Mountain for the USFS by buying it when the government could not. Let Nantahala's recorded history remember the Civilian Conservation

Corps (CCC), which helped build trails, roads, bridges, fire towers and one of the world's largest watershed research centers.

Nantahala's history records the work of community groups, volunteers and retired USFS employees clearing brush, building signs and maintaining trails with chainsaws, weed-eaters, leaf blowers and a donkey named Alf. It includes the work of volunteers with the Southern Appalachian Wilderness Stewards (SAWS), which helps maintain wilderness trails with traditional tools. It records the manual labor of the Youth Conservation Corps (YCC), the Older Americans Program and the Job Corps. Over the decades, the USFS and others have restored deteriorating structures, some built by the CCC, with painstaking detail to retain their historic integrity.

The setting for the stories of all of these people, programs and activities occurs in one place: Nantahala National Forest. The history of Nantahala National Forest, preserving those narratives, is best told by exploring its three districts, each bearing a Cherokee name: Nantahala, "land of the noonday sun"; Cheoah, "the otter place"; and Tusquitee, "where the water dogs laughed" (or "the place of house rafters, the shape of the mountains"). Let the drama begin.

PART II

NANTAHALA RANGER DISTRICT

Ten-year-old Michael found the perfect leaf near the site of the old Civilian Conservation Camp at Otto, North Carolina. In the 1930s, Ray Welch, a hungry teenager, had convinced CCC recruiters that he was old enough to join the Corps at Otto. Now, huge trees guard that historic site located in the upper watershed of the Little Tennessee River. A small limb of a large buckeye tree curved down within Michael's reach, as if to offer him a gift. The broad, oval compound leaves fanned out from leafstalks like wide, open fingers. Michael chose the widest one, folding one end toward the center. "Rip the folded part into three equal sections," instructed Ranger Rick on the National Wildlife Association webpage, "and tuck the two outer sections inside one another." Michael pinned them in place with a tiny stem. The center section of the leaf curled up like the bow of a boat. The midrib of the leaf formed the keel; its veins formed the frame. Michael set it afloat on the Little Tennessee River.

Farther west in the Nantahala Ranger District, created in 2006 by combining the Highlands and Wayah Districts, Michael could have launched his fragile craft on the headwaters of Nantahala River. There, it would have ridden the currents beneath Pickens Nose and Albert Mountain Fire Tower and through the former White Oak Bottoms logging camp. Past Rainbow Springs, the open dinghy would have slipped west of Siler and Wayah Balds and the Bartram and Appalachian Trails to float into Nantahala Lake. If it survived the dam, the tiny boat would have battled rough waters and kayakers through Nantahala Gorge to join the Little Tennessee River at Fontana Lake.

Tom Alexander (*left*) and Dan Sweetman (*right*) at Albert Mountain Fire Cabin. *National Forests of North Carolina Historic Photographs, D.H. Ramsey Library, Special Collections, University of North Carolina–Asheville.*

What if Michael had gone east in the Nantahala District to set his craft loose at the base of Whiteside Mountain near the headwaters of the Chattooga River? What if he had gone south to set it adrift on the Whitewater River near Cashiers? Could the fragile leaf blade withstand Chattooga's Class IV rapids or Whitewater's four-hundred-foot plunge? Suppose Michael had journeyed northeast to Panthertown Valley to launch his leaf on the Tuckasegee River. Would the swirling waters in the potholes of Bonus Defeat have stranded the little buckeye-boat, or would it have weathered the rocky jungle to join the Little Tennessee at Fontana?

Michael chose the Little Tennessee River, one of the most biologically diverse river basins in North America. More than one hundred species of fish live there. State or federal governments have listed about forty aquatic species as rare or endangered. In 2015, officials designated the river as a Native Fish Conservation Area. Near the USFS Coweeta Hydrologic Laboratory, Michael launched his leaf-boat to float north. Citico darters, smoky madtoms and Little Tennessee crayfish—species that exist nowhere else in the world—would have passed the small craft along its journey.

North of Highlands, the Cullasaja River joined the Little Tennessee to flow through Franklin past the ancient Nikwasi Mound. Through the Valley of Cowee, the leaf-boat joined kayakers and other drifters. In 1767, Thomas Griffiths of England had collected nearly five tons of clay in Cowee Valley to ship back to England's Wedgewood factories. In 1775, the naturalist William Bartram had explored this region and visited the Cherokees at Cowee Mound. Above the mineral- and gemstone-rich Cowee Valley and the Little Tennessee River, the Nantahala Mountains rise to the west. The Cowee Mountains rise in the east. In the 1930s, the CCC had built a fire tower there on Cowee Bald above gardens of wildflowers. Today, in nearby Alarka, exhibits along the USFS Walton Smith Interpretive Trail identify flora in a rare red spruce bog.

In Cowee Valley, the buckeye-boat passed through fourteen weirs—Cherokee V-shaped, rocky barriers that funneled fish into nets or baskets placed at each outlet. In 1881, explorers Wilbur G. Ziegler and Ben S. Crosscup wrote in *The Heart of the Alleghanies*:

> *A long dug-out* [is], *fastened to the bank at the end of a beaten path; and between the trees see a lonely cabin on the opposite side of the river.... The spot appears to be too isolated to be either pleasant or romantic. One of the many fish traps seen in all the mountain rivers is near the cabin. It is built, like they all are, in a shallow reach of the river. It consists of a low V-shaped dam, constructed of either logs or rocks, with angle pointing downstream. The volume of the water pours through the angle where is arranged a series of slats, with openings between, large enough to permit the passage of a fish into a box set below for its receptacle. Every day its owner paddles his canoe out to the angle of the dam and empties the contents of the box into the boat.*

Michael's tiny boat skimmed over schools of sicklefin redhorse, a rare species first discovered in 1992. A startled fawn, panicked by passing boaters, tumbled into view along a shrubby bank. Domestic cattle waded beside an open meadow. Introduced bamboo out-competed the native river cane, used by the Cherokees for baskets, blowguns and fishing poles. First discovered in Virginia in 1985, the rare and threatened Virginia spiraea grew along the riverbank about one-half mile below Iotla Bridge. Papaw trees have rooted near a former Cherokee village. Wide, silent wings of a great blue heron

The Cowee Bald Observatory, built in 1933 by the Civilian Conservation Corps, allowed excellent views of the Nantahala, Great Smoky, Plott Balsam and Great Balsam Mountains. *National Forests of North Carolina Historic Photographs, D.H. Ramsey Library, Special Collections, University of North Carolina–Asheville.*

Kayakers paddle the Little Tennessee River with the Highlands Biological Station. *Photo by author.*

lifted him further downstream to an undisturbed fishing spot. Overhead, the unmistakable, odd call of a yellow-billed cuckoo echoed from the dense canopy. Kingfishers rattled and gnatcatchers chattered while mother mallard and ducklings curved back to a secluded safe haven.

Beneath a Piper Cub airplane following the course of the Little Tennessee River, Michael's buckeye-boat drifted under draping boughs of black cherry, around a turtle sunning on a large log and a beside a streamside sycamore losing its bark, like dried skin peeling after a sunburn. The tiny craft passed under a swinging bridge and a highway to float into Fontana Lake near Almond.

WAYAH BALD/WILLIAM BARTRAM

In November 2015, an overcast, fifty-degree Saturday invited hikers to burn a few post-Thanksgiving calories. Gail Lehman, a hike leader with the Nantahala Hiking Club (NHC), carpooled hikers from Franklin along Wayah Road (NC 1310) and FSR 69/69B to the Wine Spring Bald Trailhead. With a crooked mountain laurel branch as her walking stick, Gail flipped fallen limbs from the footpath. As a hike leader and trail maintenance crewmember, she often combines both activities in the same outing. Ragged ends of a broken limb across the trail entangled a hiker's boot. Gail and a fellow hiker shoved it aside.

"I need to check that one for downfall," said Gail, spotting an old logging road below the trail. "I'll give a report to the NHC trail manager. Maybe we'll return with a maintenance crew, so we can lead a hike there."

When Gail moved to Franklin in the 1970s, she first volunteered in NNF with the Horseback Search and Rescue Team. Now she has traded hoofs for boots, leading hikers that day through the fallen, multicolored leaves of poplar, oak and maple. Twenty boots overturned the masses into an unsettled wake. Piles of dirt along the trail suggested recent visits by a foraging bear or boar—surely wild turkeys could not create such unearthed disturbances. The hikers followed a short segment of the Bartram Trail across a high-elevation meadow, maintained as a wildlife foraging field. On the open ridgeline, two camouflaged deer hunters, capped in orange toboggans, shared the trail with hikers, flagged in orange fleece. Gunshots from the USFS Dirty John's Shooting Range echoed in the valley.

Long-range western views enticed hikers to explore the Chunky Gal and Fires Creek Rim Trails in the Tusquitee Ranger District. Near Sawmill Gap, the Bartram Trail continued northwest. The hikers followed an old roadbed, instead, looping back toward the trailhead near Wine Spring Bald. Along the return route, the rounded summit of Wayah Bald dominated the leafless eastern view. Beneath the brown forest floor, spring beauties, Dutchman's breeches and squirrel corn remained dormant until the spring green season.

Wayah Bald is one of the highest mountains on the North Carolina section of the Bartram Trail. For four years, William Bartram (1739–1823), considered by many to be America's first native-born naturalist-illustrator, journeyed through eight colonies of the Southeast, documenting his botanical, geographical and cultural discoveries. In the original 1791 book, republished later under the shortened title of *Travels of William Bartram*, his scientific observations promoted increased interest in the region's flora and fauna and significantly contributed to natural history literature. In the 1979 publication of *Bartram Heritage*, the Bartram Trail Conference wrote:

> Hortus Siccus *or "dry garden" is a Latin term botanist formally used for an herbarium, a collection of dried plant specimens, carefully identified and labelled which they had assembled for the benefit of later observers. The term also aptly describes Bartram's* Travels, *which contains literary specimens of southeastern North America, dried and preserved by the process of writing…for successive generations of readers.*

As son of John Bartram, a Quaker farmer and the English colonies' first botanist, William became enamored of natural history while tending his father's 5-acre botanical garden. John Bartram owned a 105-acre farm on the Schuylkill River near Philadelphia. Commissioned as a plant collector in the forests of the New World, John sent seeds, cuttings and living plants from an estimated two hundred different American plant species to nearly fifty subscribers, including Carl Linnaeus, the queen of Sweden, dukes, earls and other European patrons.

At age twenty-one, William left his father's Philadelphia home to live with his uncle on the Cape Fear River in North Carolina. In about 1709, Indians had killed his uncle's father (William's grandfather) and captured his uncle. After receiving a ransom, the warriors released the young child (as well as his sister and mother) to return to Philadelphia. Now, as an adult, William's uncle welcomed the young botanist to help him manage a trading post on the Cape Fear River. After King George III appointed John Bartram royal botanist of the New World in 1765, William joined his father on a yearlong plant-collecting trip in South Carolina, Georgia and Florida. After the trip, William attempted to set up an indigo plantation in Florida, but it failed miserably. After years of struggle, he received sponsorship in 1772 from John Fothergill for his approximately 2,400-mile journey across the Southeast.

To memorialize Bartram's historic expedition, several organizations had formed since 1914, promoting heritage conservation projects. In 1975, governors of the eight states through which Bartram passed created the joint Bartram Trail Conference and proposed a continuous historic trail system that followed Bartram's entire route. However, a 1977 study found this objective impractical. Two centuries of human development had erased his exact footpath. The Bartram Heritage Study suggested highway markers, interpretive centers and memorial parks to preserve Bartram's history and increase public awareness. Bartram Trail proponents suggested building non-continuous trail sections along his identified route, potentially designating some as National Recreation Trails. In Louisiana, Georgia and North Carolina, portions of his journey retained natural elements experienced by Bartram. In 1977, eleven North Carolinians organized the NC Bartram Trail Society and helped the USFS build the trail. Today, Georgia's thirty-seven trail miles join North Carolina's seventy-nine miles to form a continuous route from Georgia's Russell Bridge at the Chattooga River to Cheoah Bald in Nantahala National Forest.

In 1993, board members of the NC Bartram Society met at the home of its president, Burt Kornegay, to discuss the preparation for the Biennial Bartram Society Conference. That October, the organization would host the event at Fontana Village, welcoming "fellow Bartramites from all over the country to see the splendid fall leaves of western North Carolina and hear speakers present information on natural history, the Cherokees, and William Bartram's travels." They chose Wilma Dykeman, a regional author and historian, as the keynote speaker. Local guides would offer a nature walk at Joyce Kilmer Memorial Forest, a boat trip on Fontana Lake, an auto caravan tour of Bartram's travels through Cowee Valley and a hike on the Bartram Trail near Franklin. Raising a Bartram Trail T-shirt on the slender trunk of a *Franklinia alatamaha*, a tree discovered by the Bartrams in 1765 near Georgia's Altamaha River, the board members adjourned the meeting and saluted their "flag." In 1776, William gathered seeds from the fragrant, white-blossomed "rare and elegant flowering shrub" to take back to Philadelphia. Today, the small tree no longer exists in the wild. All of its cultivated specimens, named by Bartram for the family's friend, Benjamin Franklin, are descendants

William Bartram. *Society Portrait Collection [V88], Historical Society of Pennsylvania.*

from those seeds. In 1976, a nurseryman estimated that there were only about one hundred mature *Franklinia* in the East. At the tenth annual meeting of the NC Bartram Trail Society in 1987, members received seedlings of the special tree. Where were they planted? How many survived? How big are they now? Several local nurseries now propagate *Franklinia*, and naturalist Jack Johnston grows and distributes specimens of this treasured species throughout Macon County.

At the North Carolina–Georgia border, south of Highlands, a Bartram Trail hiker heads north toward Osage Mountain Overlook along a rock wall, built by the Youth Conservation Corps in 1978. After crossing NC 106, a steep climb leads to Scaly Mountain (4,804 feet) straddling the Eastern Continental Divide. Rock outcrops and wet seepage areas provide nourishment for special plants, like twisted-hair spikemoss, Biltmore sedge, kidney-leaved grass-of-Parnassus, bog goldenrod and grass pink. Challenged by environmental elements, dwarfed oak trees thrive on rocky cliffs. Flame azaleas, rhododendron and laurels bloom in June.

The NC Bartram Trail then crosses the Fishhawk Mountains in NNF, considered one of the most scenic portions on the trail. A spur trail summits Whiterock Mountain and offers fantastic distant views. A plaque on top of Big Fishhawk Mountain pays tribute to William Bartram. Near an outcrop at Jones Knob, fringe trees wave white feathery crowns each spring. In *Literary Excursions in the Southern Appalachians*, George Ellison described meeting his first fringed beauty on Jones Knob: "Framed by four or five fringe trees in full bloom, the view over the Little Tennessee Valley toward the Nantahalas looming on the western horizon was spectacular." Bartram's Trail descends the Fishhawks into the Little Tennessee Valley and follows narrow country roads around Franklin.

Bartram himself did not go over the Fishhawks but rather followed a Cherokee traders' path and ancient trail along the Little Tennessee River from its source. In May 1775, Bartram traveled through the Little Tennessee Valley, passing through a Cherokee settlement at Echoe at the mouth of

Cartoogechaye Creek. Three miles distant, Bartram visited a Cherokee town that he called "Nucasse" (present-day Franklin). The Cherokees named this council house mound "Nikwasi," built thousands of years earlier. In 1761, before British soldiers burned the townhouse, more than five hundred Cherokees had lived near Nikwasi. The Cherokees returned and rebuilt the townhouse, but in 1776, North Carolina militia led by General Griffith Rutherford burned Nikwasi again in an effort to destroy the Middle Towns of the Cherokees. Downtown Franklin now surrounds the ancient mound, listed in the National Register of Historic Places. A North Carolina historical marker and interpretive signs to exhibits provided by the Blue Ridge National Heritage program tell its story. On Franklin's Iotla Street, a highway marker records Bartram's visit to the area. South of Franklin, an eleven-mile-long Bartram Canoe Trail (or Tanase Canoe Trail, using Bartram's name for the river) explores the Little Tennessee River between Otto and Franklin.

The Little Tennessee River flows north through Franklin and the Cowee Valley, where Bartram visited Cowee, a major settlement of the Cherokee Middle Towns. In the Piedmont of South Carolina, Cherokees had fled their Lower Towns ahead of advancing military forces. Some Cherokees came to Cowee and other Middle Towns in the Little Tennessee River Valley. Others settled in the Shooting Creek area or the Valley Towns between Andrews and Murphy. Some continued northwest of the Unicoi Mountains, now Tennessee, to the Cherokee Overhill Towns. At Cowee, Bartram found more than one hundred homes, spanning both sides of the Little Tennessee River. Extensive fields grew corn, beans and squash. A council house sat atop the ancient Cowee Mound.

Archaeological research has dated the mound from approximately AD 600. Pollen sampling from these fertile bottomlands revealed evidence of agricultural practices nearly three thousand years ago. The University of North Carolina–Chapel Hill preserves prehistoric ceramics collected at the Cowee Mound in 1965. One year after Bartram's visit, Rutherford's soldiers burned the town and townhouse. By 1820, the Hall family owned Cowee Mound and seventy surrounding acres. When family descendants chose to return the site to the Cherokee Nation in 2003, Land Trust for the Little Tennessee (now Mainspring Conservation Trust) and the NC Clean Water Management Trust Fund helped the Cherokees broker the necessary funds. On January 8, 2007, the *Cherokee Phoenix* announced, "Eastern Band Buys Cowee Mound."

Bartram described this area as the "great vale of Cowe." He wrote, "[O]ne of the most charming natural mountaneous landscapes perhaps anywhere to be seen, ridges of hills rising grand and sublimely one above and beyond another, some boldly and majestically advancing into the verdant plain, their feet bathed with the silver flood of the Tanase, whilst others far distant, veiled in blue mists." After vividly recording the Cherokee maidens in the strawberry fields north of the settlement, Bartram headed north toward present-day Wayah Bald.

En route to Wayah Bald, Bartram Trail hikers rejoin the footpath at Wallace Branch, about two miles west of the Nantahala Ranger Station, and begin a strenuous ascent up Trimont Ridge. The trail reaches a large outcrop, named William's Pulpit in honor of Reverend William Haselden, an avid Bartram Trail Society volunteer. Between April and October, a tiny plant growing in the thin, sandy soils of the Pulpit opens its flowers from 3:00 p.m. until dusk. Even though a single plant may produce more than one hundred flowers, each tiny purple bloom of the quill fameflower opens for only one day.

Along Trimont Ridge in the Nantahala Mountains, whiteleaf sunflowers border the Bartram Trail through wildlife foraging fields. Bartram called the Nantahalas the Jore Mountains after a Cherokee village located on Iotla Creek. A well-repeated Bartram quote comes from his journeys in the Nantahalas: "[T]he Jore Mountains which I at length accomplished, and rested on the most elevated peak; from whence I beheld with rapture and astonishment, a sublimely awful scene of power and magnificence, a world of mountains piled upon mountains."

On Wayah Bald, the Bartram Trail joins the A.T. for two and a half miles. At the summit, the Wayah Bald Lookout Tower permits grand 360-degree views. USFS exhibits identify Standing Indian Mountain, Whiteside Mountain, the Cowee Mountains and many others. North of Wayah Bald, the A.T. follows the crest of the Nantahala Mountains, dips to Burningtown Gap and climbs past Tellico and Wesser Balds before entering Nantahala Gorge. The southbound A.T. hiker departs the Bartram Trail near Wine Spring Bald, crosses FSR 69 and Wayah Road (SR 1310) and heads to Siler Bald. Bartram hikers continue west over McDonald Ridge, dip into Sawmill Gap and arrive at Nantahala Lake.

Completed in 1942, Nantahala Power and Light, a subsidiary of Aluminum Company of America (Alcoa), dammed the Nantahala River to generate hydroelectric power. Started in 1929 but delayed by the Great Depression, the dam (now owned by Duke Energy) creates a deep, cold-water mountain lake that rests at three thousand feet of elevation. In preparation for the power plant reservoir, workers removed schools, stores and other buildings. They relocated bridges spanning the Nantahala River. Exhumed bodies in a local cemetery now rest at Little Choga and Aquone Cemeteries. Homesteaders, farmers and church congregations found new homes. Loggers deserted lumber mills and railway systems, leaving a train wreck at Horseshoe Bend in its aquatic grave. Lake waters submerged or displaced communities like Briartown, Little Choga, Kyle, Otter Creek and Beechertown.

Also flooded was the old town of Aquone (Cherokee for "by the river"). Cherokees occupied this region before white settlers moved in. Local historians report that in the 1830s, Cherokee natives had ambushed General Winfield Scott's troops, arriving to remove Cherokee families from their homeland. One night, in retaliation, Scott's men climbed the steep ridge between Wine Spring and Wayah Balds and attacked the Cherokee village. Soldiers herded the captured Cherokee families along Nantahala River to the detainment fort, Camp Scott, located at the mouth of Wine Spring Creek. Eventually, the military marched them over the mountains to Fort Delaney at Valleytown (now Andrews) and then on to Fort Butler at present-day Murphy.

Camp Scott became the town of Aquone. According to Francis Harper in his 1958 book, *The Travels of William Bartram: Naturalist's Edition*, Bartram most likely crossed the Nantahala River here. Almost one hundred years after Bartram's visit, more than one hundred residents occupied the Aquone community, a major stagecoach stop along the Great Western Turnpike from Asheville to Murphy.

Nimrod Simpson Jarrett lived on a large farm in Aquone. One of the wealthiest landowners in Western North Carolina, Jarrett owned thousands of Haywood, Macon and Swain County acres over his lifetime. He earned a living as a land speculator, farmer, ginseng trader and owner of talc and mica mines. His property stretched from about one mile upstream of Nantahala's juncture with Jarretts Creek (named for him) to the current Ferebee Memorial Recreation Area in Nantahala Gorge. Jarrett Bald on the Bartram Trail is also named for him. In 1855, when Jarrett's home in Aquone had burned, killing his youngest daughter, he and his family moved about six miles downstream to Apple Tree, now a USFS group campground. From Apple Tree to Nantahala Gorge, he built a ten-mile road through the rugged terrain, now part of US 19.

In the late 1870s, Jarrett departed Apple Tree on horseback to attend a meeting in Franklin and forded the Nantahala River. Midstream, a thief shot and killed him. Locals immediately launched a posse. Legends say that searchers found a footprint with a defective heel that matched the boot print of a member of the posse party. The court system found Bayless Henderson guilty and ordered that he hang—the only man ever hanged in Macon County.

In the 1930s, the Cherokee stockade, Camp Scott, also became the site of the CCC Camp Winfield Scott. The enrollees helped the USFS grade roads, fight fires and build fire towers. Captain Phillips of Fort McPherson assigned William A. Ogletree to equip the encampment with radio communications. As radio dispatcher, Ogletree remained there for six weeks until a new graduate could replace him. However, during his service at Aquone on the banks of the Nantahala River, Ogletree fell in love.

At the blacksmith shop beside the Aquone Post Office, he met a sixteen-year old, beautiful "barefoot lass." During his free time, the pair fished and swam in Nantahala River, climbed Wayah Bald and attended church. Five days a week, they took a blanket roll and a picnic lunch, packed by the CCC camp cook, into the mountains. As an eighteen-year-old recruit, Ogletree knew that marriage meant automatic discharge from the Corps. When Phillips offered him an opportunity to study radio engineering at Georgia Tech and become the chief radio operator in Atlanta, Ogletree knew that the promotion to chief would allow him to marry. The response of the "barefoot lass," recorded in Ogletree's essay, "Making of a Radioman," surprised him:

> *By now you ought to know that I will marry you—but I won't live in Atlanta. Look around! In this forest about 200 feet above us is a great spring with the best water in Aquone. Down there, about 900 feet below, is the Nantahala River. To the right and to the left as far as you can see, is the Nantahala Valley. God made no other place on earth like this one, and it is mine. When I was only 10 years old, Pa gave me this land to me and promised that someday he'd build a cabin here on this land bench for me and my husband to raise our family. Pa owns all the private land this side of the river, up to the ridge topping this mountain. Everybody else sold out to the Nantahala National Forest, but Pa would not sell it.... There's a whole forest of black walnut trees in the valley, and a grove of black cherry that my grandpa planted.*

Her mother warned her that this CCC man, William Ogletree, was too ambitious to settle in Aquone. The young couple stood quietly together "looking at the green forest and the blue river below" and then parted ways.

When Nantahala Power and Light Company began moving in, landowners around Aquone moved away. Six miles downstream from the dam, the company constructed a power station at Beechertown. Duke Energy now owns and maintains the dam and about thirty miles of shoreline. NNF surrounds much of the 1,600-acre reservoir enjoyed by boaters, swimmers and fishermen. Anglers not only fish for trout, bass, walleye and crappie but also hook descendants of kokanee salmon stocked in the lake in the 1960s by the North Carolina Wildlife Resources Commission (NCWRC). Fireworks usually brighten the skies each Fourth of July above an island near the mouth of Wine Spring Creek.

The Bartram Trail hikers catch a view of the Nantahala Dam, cross the Nantahala River, follow it downstream toward Junaluska Road (SR 1401) and arrive at the USFS Appletree Group Campground. Built in the 1960s for a Boy Scout Camporee, campground trails—like Appletree, Junaluska Gap and Diamond Valley Trails—connect to the Bartram Trail. Old apple trees still offer fruit in late summer or fall. In the 1970s, the U.S. Army Green Berets and Special Forces camped at Appletree. Wayah Bald provided radio communications for training maneuvers. At Nantahala Lake, pilots targeted the dam in simulated bombing runs. With only their heads above the water's surface, the servicemen practiced swimming without leaving a wake. Beside a lakeside diner, pilots landed Chinook and Huey helicopters at lunchtime and handed out military can openers, patches and other souvenirs to local children.

For training operations in the mid-1980s, U.S. marines from Camp Lejeune landed in the thirty-acre field at Appletree Campground. Camping for several days, they learned mountain survival techniques, like eating bugs, snakes and other available delicacies. John, the USFS campground host, and his grandson lived in a small trailer on site. When the troops left for mountain training one day, two marines remained at camp to guard the helicopters. Unsupervised and thirsty, they made a bad decision. They drove to town for beer, taking John's grandson with them. At 6:00 a.m. the next morning, District Ranger Lewis Kearney received an angry call from his campground host. Kearney immediately called Camp Lejeune.

"I want to speak to the Base Commander," he told the switchboard operator. She transferred the call to a corporal. The corporal passed him to a lieutenant. After speaking to a captain, a major and a colonel, Kearney became irritated. He told the officer that he would be reporting the story to an Asheville newspaper of how marines at Appletree involved a twelve-year-old boy in their misdeeds.

"I'll send you the newspaper clippings," Kearney said. The colonel promptly transferred his call to the general's office. A booming military voice came on the phone. Kearney thought about hanging up.

"I will see you in five hours at your camp," roared the general. "Be there!"

Kearney drove to Appletree. He explained to John that they were expecting a few guests. Soon, faint rumbles of rotor blades grew louder. Huey helicopters hovered overhead before landing in the rain-soaked ground, sinking their skids into the soft mud.

"The ramps came down and there was no problem telling who the General was," said Kearney. "I stuck my chest out as best as I could and tried to look puffy in my uniform. After the formalities and apologies to me, we marched to the tent where the two guilty parties were standing, at attention, of course; never in their wildest dreams did they think the Base Commander would be there. One poor soul immediately fainted. After the big speech and more apologies, the entourage climbed back in their helicopters and flew away. That was the day that the Brigadier Commander came to our little campground."

After leaving Appletree Campground, the Bartram Trail crests Rattlesnake Knob and descends to Nantahala River at the head of the gorge. After passing a power station surge tank, transmission lines, mountainside penstock and a discharge chute, the trail travels past the power plant at Beechertown to the USFS Nantahala River Launch site. For 1.2 miles, hikers join bikers on the paved Nantahala Rails-to-Trails Bikeway. A portion of this section retraces an old spur of Southern Railway's Murphy Branch that once transported building supplies to the power plant.

On Bartram Day, May 29, 1973, celebrating the annual event honoring William Bartram, seven Bartramites hiked this "Section Six" of the trail, carrying brand-new maps published by the NC Bartram Trail Society. With the help of students from Green Mountain Valley School in Vermont, society volunteers had recently completed this section. Dan Pitillo and the hikers identified rare plants, such as Goldie's wood fern and glade fern, as well as ginseng, umbrella leaf and other flora. They passed the Bartram bridge construction crew building two twenty-foot log bridges at the nearby Piercy Creek Trail.

At US 19/74, the Bartram Trail crosses Nantahala River east of a highway marker commemorating Bartram's meeting with Attakullakulla, the great chief of the Cherokees. In 1730, as a young man, Attakullakulla and six other Cherokees traveled to England as guests of Sir Alexander Cuming. Entertained by the king of England and painted by Hogarth in British attire, Attakullakulla and the Cherokees pledged allegiance to the British against the French and the Creek Indians. In August 1760, Attakullakulla saved the life of Captain John Stuart during the fall of Fort Loudoun (near present-day Knoxville). When he met Bartram in 1775, Attakullakulla was on his way from the Overhill town of Chota to Charleston to see John Stuart, now superintendent of Indian affairs. Bartram wrote:

> *After crossing this large branch of the Tanase, I observed, descending the heights at some distance, a company of Indians, all well mounted on horse-back; they came rapidly forward: on their nearer approach, I observed a chief at the head of the caravan, and apprehending him to be the Little Carpenter, emperor or grand chief of the Cherokees, as they came up I turned off from the path to make way, in token of respect...his highness with a gracious and cheerful smile came up to me, and clapping his hand on his breast, offered it to me, saying, I am Ata-cul-culla.*

Bartram only traveled a few miles farther. He never made it to his destination, the Cherokee Overhill Towns. Attakullakulla warned him, accurately, that trouble was coming. During the bitter pre–Revolutionary War days, Bartram turned around—some believe at a place now known as Sweetgum in Graham County—and returned to Cowee.

Past the Nantahala River, the Bartram Trail begins a 3,000-foot climb out of Nantahala Gorge, crosses Ledbetter Creek multiple times and passes chutes, slides and half a dozen waterfalls, including one named "Bartram." About five miles later, hikers join the A.T. for the final two-tenths of a mile to Cheoah Bald (5,062 feet), the northern terminus of the Bartram Trail.

Wayah Bald Lookout Tower provides a view of the mountains that embraces the northern and southern ends of the Bartram Trail—south to Rabun Bald in Georgia and north to Cheoah Bald, in the opposite direction. The panoramic views on clear days prompted the USFS to build its fire tower atop Wayah Bald (5,324 feet) instead of Wine Spring Bald (5,440 feet), the highest mountain on Nantahala Ridge. In 1929, USFS Gilmer Setzer helped build the oak and chestnut structure and served as its first fire warden. His son, Ed, often joined him overnight in the tower's cabin. His uncomfortable trundle bed slid under his father's cot to increase daytime working space.

"When the wind blew," said Ed, in an interview with former USFS interpretive specialist Bill Lea, "the wooden tower would move and creak." On most days, Ed's father walked home for lunch at the Wilson Lick Ranger Station, where Gilmer and his bride had honeymooned years earlier. The couple's first daughter, Mary, was born there. Once, when Gilmer only had a wet match to start a fire in Wilson Lick's fireplace, he broke open a few

Bill Canon, a Wayah Bald fire warden, pinpoints a fire's location using an adelaide firefinder. *Bill Lea.*

shotgun shells and placed them on dry kindling. A blast from his gun ignited the pile and singed his eyebrows, but the Wilson Lick cabin still stands.

Over time, severe, high-elevation weather deteriorated the wooden lookout tower. In a 1995 article of the *North Georgia Journal*, Richard Byrd, son of Zeke Byrd, an early NNF ranger, described a garden club's frightening experience. While hiking a short section of the A.T. across Wayah Bald, the ladies sought shelter in the tower during a thunderstorm. USFS staff working nearby heard the old floor groaning under their weight.

"Ladies, get out of there!" yelled Grady Waldroop. "The floor is about to fall!" Too petrified to move, the ladies waited for Waldroop to carry them down the ladder one at a time. Shortly thereafter, the USFS demolished the aging structure.

After improving the two-and-a-half-mile Wayah Bald access road in 1934, the CCC replaced the wooden tower with a fifty-three-foot tower of rock and timber. The corpsmen built it next to a USFS ranger's cabin moved to Wayah Bald in 1927. Since the late 1880s, the bald had become a popular destination for outings and church services, so the men built a deluxe model.

The top deck provided equipment and living quarters for the fire wardens. On the second level, an observation deck offered scenic views for visitors who climbed an oak staircase to an enclosed platform. Twenty-four feet above the ground, visitors enjoyed views through large casement windows built into the tower. Along a rustic balcony, benches provided seating outside. Oak shakes covered its triangular roof. Flagstone terracing at its base joined native flame azaleas to blend the man-made tower with the rocky summit.

On September 6, 1937, four hundred ceremony guests celebrated the new tower, dedicated to NNF district ranger John B. Byrne. In 1934, at age thirty-three, he had died, reportedly, from pulmonary complications following poisonous gas exposure in World War I. During his career, Byrne helped establish about six CCC camps under President Roosevelt's national plan for economic recovery. Reverend J.A. Flanagen of Franklin paid tribute: "We pray this tower may be dedicated to the memory of one who we honored… [standing] here amidst the clouds and the mists, the sunshine and the rain."

In 1945, the Forest Service decommissioned the Wayah Bald Lookout Tower for fire detection. After removing the fire warden's cab, the balcony, an interior oak staircase and eighteen feet of rock, they capped the rock base with a concrete platform. Rock masons recycled the removed rocks into an exterior stairway on the side of the tower. However, by the 1970s, age, weather and neglect had weakened its walls and cracked the mortar.

In the 1980s, some suggested that the USFS demolish the crumbling structure and replace it with an observation platform. The Forest Service, instead, preferred to preserve its historic character. Ranger Kearney explained to the *Franklin Press*, "The Forest Service has let too many historic treasures get away from it in the past. Plus there's a lot of local interest in fixing up the old tower." Assistant District Ranger Bill Sweet agreed: "We knew if we tore it down and replaced it we would be losing something that has great historical significance as far as Nantahala National Forest is concerned."

The agency hired employees with the Older Americans Program to replace the top three feet of rock and create a covered, open-air viewing platform. They built a new roof of hemlock beams, locust rafters and cedar shakes. NNF provided the rocks and lumber, and $1,500 in funds purchased the concrete and cedar shakes. Men from the Lyndon B. Johnson Job Corps Center on Wayah Road built new restrooms. The Pisgah National Forest Job Corps built exhibits.

In 2009, teenagers tested homemade bombs on the Wayah Bald Tower. Six times, the young men threw Molotov cocktails and explosives into the

In 1935, the CCC built the elaborate Wayah Bald Fire Tower beside a cabin that had been moved to the summit by the USFS in 1927. *National Forests of North Carolina Historic Photographs, D.H. Ramsey Library, Special Collections, University of North Carolina–Asheville.*

bottom level of the tower, igniting but not collapsing the stone structure. Coincidentally, during a routine road check, the USFS discovered bomb-making materials and a broken USFS padlock in their trunk. Law enforcement officials arrested them for making weapons of mass destruction.

By 2010, the USFS had noticed deteriorating cracks in one of the walls of the tower. With the aid of $75,000 from the Federal Economic Stimulus Package, the USFS hired contractors to make the repairs. Before tearing down the exterior rock stairway, the contractor mapped the wall's image and designated a number for each rock's placement. Crews then numbered the rocks one by one as they dismantled it. After completing the structural repairs, each rock returned to its historic placement, using river sand to mix the mortar, as the original builders had.

Sunset from Wayah Bald is a soothing experience. As the sun slips below the horizon, its weakening strength no longer squeezes a squint or casts a shadow. As if christened with shiny gold dust, its retiring glow illuminates mountain ridges. Backlit with a golden flood of light, silhouettes of tall trees stand in bold relief on distant mountains. Lining the ridgelines, dark tree forms look like porcupine quills fending off predators. Across the crest of some mountains, faint yellow tones blend smoothly into the domed outlines.

Motoring on Wayah Road. *Macon County Historical Society and Museum.*

As if gently edged by a Renaissance artist, curved terrestrial boundaries between heaven and earth soften into one harmonious portrait.

A quiet return drive along FSR 69 bursts into action; quick reflexes brake for a ruffed grouse flushed by the roadside. Another brake reveals the rounded, shadowy form of a barred owl scouting for a late meal—a reminder that the day is not over yet. Come nightfall, Nantahala's forests will support the less celebrated nocturnal species. For now, in the soft, diffused late June twilight, multitudes of summertime wildflowers ripple by the roadside, like goatsbeard, fire pink, columbine, Bowman's root and purple-flowered raspberry. Extending deep into the forest understory, the orange and deeper-orange, yellow and paler-yellow azaleas paint the hillsides along the graveled road. Thousands of blossoms share colorful pigments—a yellow highlights an orange, and an orange streaks a yellow. Pretty pinks blush and pure whites shine, complementing nature's tree-sized bouquets of orange azaleas. Rosy cups of mountain laurel mix with the natives.

Most of the roadside azaleas are types of *Rhododendron calendulaceum.* On Wayah's summit, a different species of azalea joins the group. White or smooth azaleas (*R. arborescens*) gather at the base of the tower and line the visitor's trail. There, slight breezes perfume the forest with the sweet fragrance of smooth azalea, luring pollinators and human visitors to approach for a closer look. Elegant pale-pink and pure white flowers advertise pollen-rich, deep-red stamens. Constant catbird chatter overwhelms a towhee's unhurried, casual chant. Like a bugler playing "Taps" signaling the end of a day, a veery sings a magical, flute-like tune in the distance.

WILSON LICK RANGER STATION

In 1912, the USFS purchased the Wayah Bald tract from the Macon Lumber Company. This purchase included Wilson Lick, supposedly named for Mr. Wilson, who owned a cabin and provided salt blocks for free-ranging cattle here. In 1916, Wilson Lick became the first ranger station for the impending Nantahala National Forest.

With five rooms and a loft, the cabin served as home and office for Nantahala's first ranger, Grady Siler. In a 1980 interview with Bill Lea, Siler's widow, Mary Wilbanks Siler, described her first visit to Wilson Lick. Spending $500 for weekend supplies, including a $10 rental fee for each of the six cars and a truck, Grady drove to the ranger station with family and friends on June 15, 1919. Stopping hourly to drain and refill car radiators with cool water, the parade of Model Ts arrived at Wilson Lick after a daylong trip from Franklin.

From wood stove to ceiling, a stovepipe sent smoke skyward. Today's brick chimney was added later. Kitchen cabinets, called "food safes," were closed with wire-screen doors. In a room beside the kitchen, a long table could seat ten people. Fresh water arrived to the back of the cabin from a spring two hundred yards away by an innovative pulley and cable system.

"When a thump was felt," wrote Richard Byrd, "slack was let out of the cable and the bucket dipped into the spring. Then it was carefully reeled back up the hill to the station." Free-roaming hogs were kept out of the ranger's cabin by a fifty-foot fence surrounding the cabin. During the year and a half that Ranger Siler lived at Wilson Lick, he rode a horse and motorcycle on trails and primitive roads to make his rounds in NNF.

Wilson Lick Ranger Station. *National Forests of North Carolina Historic Photographs, D.H. Ramsey Library, Special Collections, University of North Carolina–Asheville.*

Between 1929 and 1930, the USFS planted a one-acre Norway and red spruce experimental forest behind Wilson Lick to replicate the Black Forest in Germany. The agency purchased the foot-tall seedlings from Champion Fibre Company in Canton. In about 1988, southern pine beetles destroyed the evergreens, so lumbermen harvested the wood in a salvage sale.

In the 1930s, Wilson Lick became a base camp for Camp Arrowood CCC corpsmen, who helped to build the Wayah Bald Road and fire tower. They replaced chestnut bark–covered walls and shutters on the ranger's cabin with newly hand-riven white oak shakes. They constructed beds in the open loft and made several bunk beds downstairs, placing some in the living room. The USFS officer in charge of CCC projects occupied the front room as an office. Wilson Lick also served as a communications center. The station's phone, telephone lines and wire service connected it not only to Washington, D.C., but also to telephone service in Franklin, Wayah Depot and the fire towers on Wayah Bald, Cowee Bald and Standing Indian Mountain.

Between 1938 and 1939, NCWRC's game refuge manager, George Crawford, lived at Wilson Lick with his wife and two sons. Although the pulley system was intact, Crawford collected fresh water at the springhead because "the 'lazy gal' was too hard to crank," he reported in an interview. On occasion, the family ate dinner in the car while waiting for a skunk to leave the cabin. Once, when someone shot a skunk in the kitchen, "it always smelled like a polecat on cloudy days." Like Gilmer Setzer and his bride, Ed Waldroop and his wife also spent their honeymoon at Wilson Lick. From 1948 to 1949, Waldroop served as warden for the Wayah Game Refuge and was the last resident at the historic ranger station.

Methodist volunteers from Franklin repaired the cabin for a Youth Fellowship Retreat. The young people slept in bunk beds constructed by the CCC. Men and women used separate outhouses. A porch and a 168-square-foot garage were later added but have now been removed.

Between the 1940s and the 1960s, the USFS held annual first-aid and fire training camps each fall at Wilson Lick. The Forest Service hired an on-site cook. At the week's end, spouses joined the firefighters for a festive celebration. A 1966 article entitled "First Ranger Station Still Stands" described the event: "To the accompaniment of banjo and fiddle, every man had to do a solo dance according to his best interpretation of the music."

By the late 1960s and early 1970s, Wilson Lick had fallen into disrepair. Employees with the Older Americans Program helped restore it. In addition to other upgrades, they made new wooden shakes for the roof and sides of the cabin. Two crews of six men from the Arrowood Civilian Conservation Corps (or Arrowood Job Corps Center) worked for five and a half months on alternate weekly shifts to help rescue Wilson Lick. Located on the site of the former CCC encampment, the Arrowood Job Corps was the first USFS Job Corps center to open in the United States.

Now known as the Lyndon B. Johnson Job Corps Center, the site opened in February 1965. "The first contingent of young men to be enrolled… arrived at the rural conservation center in the Nantahala National Forest," reported John Parris of the *Asheville Citizen*, "carrying their belongings in suitcases and paper shopping bags from the city streets and rural pockets of poverty." Classrooms at the center, previously occupied by North Carolina State College summer forestry students, now held corpsman studying vocational training in welding, plumbing, woodworking and automotive mechanics. Instructors also taught basic courses of math and English. Some of the old CCC buildings were remodeled, but administrative buildings and a dining hall required new construction. The staff included career USFS employees and school educators. Former Colorado State University forestry instructor H.R. Price acted as camp director. Jack S. Kelley, former assistant ranger for Pisgah National Forest, served as deputy

"On April 4, 1966, I was eighteen and began my career with the USFS, making $1.62 an hour," said Tim Southards. "Many of my early supervisors were men of the 'Greatest Generation,' who had served in World War II. I had a long and wonderful career with the USFS." Pictured is Southards (*first from left*), wearing a hard hat, attending the 1967 fire school at Wilson Lick Ranger Station. The box in the photograph, per retired USFS employee Adam Henry, is a climate station with three thermometers in it—a standard, a minimum and a maximum-read thermometer. It also contains a paper-charting device that records the humidity for a week at a time. Several are still used at the Coweeta Hydrologic Lab. *Tim Southards.*

director of work projects. In the 1970s, restoration of the Wilson Lick cabin became a primary focus.

Splitting shakes with maul and froe (a shake axe), like the men of the CCC, the Job Corpsmen fashioned large shingles for the exterior of the cabin and smaller ones for the roof. Severe weather at the high elevation halted wintertime progress, but the next year, a local newspaper reported, "They have completed a faithful restoration of the historic ranger station."

In June 1983, the USFS opened Wilson Lick to the public for the first time since 1949. Ranger Kearney asked the son of Zeke Byrd, the second ranger stationed at Wilson Lick (1926–32), to be its volunteer USFS interpreter of NNF history on summer weekends. "My job [was] to interpret forest history and to relate my experiences as a young boy with my father while he was head ranger" explained Richard, an engineer for a television station in Atlanta. "It's a task I approached with relish each season."

Richard and Zeke Byrd, Mr. Wilson, grazing cattle, honeymooners, wildlife officers, the CCC and the Methodists, as well as the fire wardens, telephone lines and fire training camps, are all now part of Wilson Lick's history. A visit today, though, is a return to the past. The "lazy gal" is gone, but the spring still percolates cool mountain water. The original ranger station still dominates the open field. Giant old trees crown the edges guarding untold stories of Wilson's past.

REVEREND ALBERT RUFUS MORGAN

Graveled FSR 69 ends at NC 1310, part of the sixty-one-mile Mountain Waters Scenic Byway through NNF. A right turn onto NC 1310 travels past Nantahala Lake, Bartram Trail and Nantahala River and ends at Almond near Fontana Lake. A left turn onto the scenic byway leads to Franklin and Highlands. At FSR 388, a two-mile side trip on a graveled road leads to the A. Rufus Morgan Trailhead. Constructed by the USFS and maintained by Nantahala Hiking Club, the trail was dedicated on October 15, 1983, in memory of Reverend Albert Rufus Morgan. That day would have been his ninety-eighth birthday. He had passed away on Valentine's Day earlier that year.

Four years earlier, USFS officials had proposed a different route for the A. Rufus Morgan Trail. Along Rough Fork from its juncture with FSR 388 to its headwaters near Siler Bald, the Forest Service had planned to intersect this footpath with the A.T. That year, on Morgan's ninety-fourth birthday, USFS Arch Nichols, the Appalachian Trail Conservancy (ATC), Morgan himself and two dozen friends gathered at the proposed trailhead. The North Carolina USFS supervisor, George Olson, handed him a plaque and said, "I hope that everyone, when they hike that trail, will be reminded of your trail leadership....He has pioneered public interest in trails at a time when the Forest Service thought of trails as little more than access ways. Reverend Morgan was busy explaining to us that trails have other values. He was ahead of us."

Arch Nichols recognized Morgan's single-handed upkeep of fifty-five miles of the A.T. and his twenty-year service on the ATC Board of Managers. However, the Morgan Trail to Siler Bald was never completed.

Today, instead, A. Rufus Morgan Trail is a moderate, one-mile loop trail up the Left Prong of Rough Fork to a beautiful sixty- to seventy-foot waterfall. Tucked into a quiet hardwood forest, wildflower gardens bloom each spring. When the trail first opened, Assistant Ranger Bill Sweet said, "Dr. Morgan was a conservationist and an environmentalist before the words were even coined. We need people like him—he inspired us."

In the early 1800s, Morgan's great-grandfather, William Siler, moved from Asheville to Macon County. His brothers—John, Jacob and Jesse—also came. Jacob Siler and William Brittain established a store near the mouth of Wayah Creek. His brother Jesse accumulated nearly one thousand acres and settled in the current town of Franklin. In the Smoky Mountains, Jesse grazed livestock on a grassy bald that now bears his name. Silers Bald (with an *s*) lies on the A.T. west of Clingmans Dome. William Siler, who owned most of Cartoogechaye Valley, about seven miles west of Franklin, is honored by Siler Bald (no *s*), on the A.T. south of Wayah Bald. Albert Mountain, on the A.T. near the USFS Coweeta Hydrologic Laboratory, bears the name of Morgan's grandfather Albert Siler.

In the midst of Cherokee natives in the Valley of Cartoogechaye, William Siler built his home in about 1838. When soldiers forced Cherokee natives west to Oklahoma, the Silers lost many Cherokee friends. One local Cherokee chief, Chuttahsotee, and his wife, Cunstagih, escaped and returned to the Cherokee Sandtown settlement in Cartoogechaye. Siler deeded land to the chief. As landowners, the chief and his wife could legally stay. William's son, Albert Siler, and his wife, Joanna Chapman, donated two acres of land for the construction of St. John's Episcopal Church. At Chief Chuttahsotee's death in 1878, Reverend John A. Deal held the first burial service at St. John's. In 1881, the union of Albert Siler's daughter Fannie to Alfred Morgan was St. John's first wedding. In 1885, Reverend Deal baptized their fourth of nine children, Albert Rufus Morgan.

In his memoir, *From Cabin to Cabin*, Rufus Morgan recalled his early youth "in a log cabin under the shadows of ridges of the Nantahala Mountains where grow the shooting star, arbutus, hepatica and bloodroot." At age six, his family moved to Murphy. After completing high school in Waynesville, Morgan graduated from the University of North Carolina–Chapel Hill and the General Theological Seminary in New York City and attended Columbia University. Soon after he married Madeline Prentiss, the couple moved to Penland, North Carolina, where Reverend Morgan became headmaster of an Episcopal mission school and ministered to churches in Avery, Yancey and Mitchell Counties. On the campus of the Appalachian School, he enlisted the work of his sister, Lucy Morgan, in the 1920s, to start the Penland School of Handicrafts, helping rural women augment family incomes while preserving traditional artistry.

Later, Morgan moved to South Carolina. In 1925, he learned that the Episcopal Church had abandoned his beloved St. John's Episcopal Church, and he became homesick. When the church offered him a position in Franklin in the 1940s, he readily accepted and served as a circuit preacher to ten area churches. His highest priority, though, was rebuilding St. John's.

Local citizens had moved the graves of family members to a cemetery in Franklin, leaving gravestones scattered in the deserted church courtyard. They had feared that over the years, encroaching overgrowth would consume the neglected plots. When Morgan learned that the grave sites of his grandparents had been moved, he refused the relocation of the grave sites of his mother and infant twin sisters. His grandmother's grave site marker, cast aside during the exodus, became the front cornerstone for the new St. John's Episcopal Church.

Nantahala National Forest donated lumber for its restoration. At the time, the USFS often provided wood for the construction of country churches. Nantahala's logs structured the framework and weatherboarding. A small rented sawmill cut shingles for the roof. Mature white pines in the churchyard became paneling. After finding St. John's old bell, Rufus Morgan fashioned a separate, open-framed bell tower out front. Reopened in 1945, St. John's Episcopal Church of Cartoogechaye continues to hold services today.

Reverend Albert Rufus Morgan. *Albert Rufus Morgan III.*

In 1946, Morgan was one of three founding members of the Macon County Historical Society, formed to save the Nikwasi Indian Mound. It raised $1,500 to purchase it from Roy Carpenter. In 1947, Morgan helped build community center cabins on property he gave to St. John's Church. One cabin now serves as the clubhouse for the Nantahala Hiking

Club. In a cabin nearby, the Nonah Weavers, under the direction of Frances Barr Cargill, have also preserved mountain artistry as the largest, active hand-weavers group in the South. During the Great Depression, educational opportunities were limited for young girls. Morgan's niece, Frances, spent summers learning the traditional skills at the Penland School of Handicrafts. After graduating from Berea College and attending George Washington University, she moved to Macon County when Uncle Rufus asked her to lead the Nonah Weavers project. When Frances married a priest and moved to New England, fellow Penland student Sally Kesler directed the program.

In November 2015, a visit to the Nonah Weavers found the gracious, soft-spoken Frances Cargill busy at her loom. Ninety-five-year-old fingers deftly threaded a shuttle of fibers between longitudinal warp yarns held taut by a large wooden loom. The rhythm of half a dozen clacking looms echoed memories of past generations. Each predesigned work of art followed a numbered, woven pattern, explained weaver Kathy Slagle Tinsley, a Siler descendant. Like Frances and Sally, Kathy's grandmother Louise Arthur Slagle attended Penland's School and was one of Nonah's original weavers. When artisans assigned a number to each letter of the alphabet, said Kathy, the final textile product could spell a hidden message. A framed, handwoven cloth designed by the Nonah Weavers hangs in the living room of Morgan's grandson. Its interwoven pattern pays a permanent tribute to Albert Rufus Morgan.

Sally Kesler was "a very efficient and talented craftswoman," Morgan wrote, "and the most knowledgeable of our hikers." In 2015, ninety-year-old Sally returned from St. John's Church and entered the Nantahala Hiking Club Cabin to discuss her forty-year friendship with Reverend Morgan. Upstairs, she lived at the Cartoogechaye Community Center, surrounded by books, papers and her silkscreen artwork.

"Sorry I'm late," she said, leaning on her walking cane. "I was planting shrubs at St. John's with a resident PhD—a professional hole digger."

Sally spoke of Morgan's love for the natural environment. He not only single-handedly maintained the A.T. in NNF from 1941 to 1968, but he, Sally and Frances also founded the Nantahala Hiking Club (NHC) in 1968. Hiking was a passion for Morgan. He hiked all the trails in the Great Smokies. Mount Le Conte was his favorite destination. From 1928 to 1977, he climbed Mount Le Conte 172 times. In 1973, three weeks before his eighty-ninth Le Conte birthday hike, he hiked up Le Conte on the Trillium Gap Trail to officiate a sunset wedding at Cliff Tops for two University of

Pennsylvania medical students. With poor hearing and eyesight, he made one of his last Le Conte climbs for his ninety-second birthday.

"I never went to Le Conte if Dr. Morgan wasn't going," said Sally with a nostalgic glow, "and I never missed his annual blueberry hike to Wine Spring Bald. It was the Blueberry Yum-Yum Pie that brought the NHC together! Dr. Morgan froze the leftovers in case the berry crop failed one season."

As a longtime club member, Sally proudly stated, "Nantahala Hiking Club is a vital part of NNF. We have done a lot to help the Forest Service." She has led hikes, marked trails and scouted out new routes. She also was a professional A.T. trailblazer, repainting faded trail blazes.

"You had to know which tree bark held white paint the longest," she said, "and you couldn't make drippy corners! I always made perfectly square corners. We tried to teach a member of the Bartram Trail Society how to blaze, but he painted from a distance. He didn't stand close enough to the tree. He didn't take it seriously. His corners dripped!...Drippy corners are made from afar!"

In an interview with *Foxfire*, Morgan described a scouting hike assigned for Sally to lead. Local advocates of the new Bartram Trail wanted to determine the route most likely taken by William Bartram two hundred years earlier in the Franklin area. Sally was familiar with the old spur trails off the A.T. near Wine Spring Bald. After a late start, the group stumbled along an obscure, overgrown trail. Nearly blind, the eighty-eight-year-old Rufus Morgan lost his footing and fell in a creek, hitting his head. Darkness closed in, delivering a cold October evening. Without food, water or warm clothing, the five hikers built a fire and spent the night. The next morning, frantic family members met them at the end of the trail near Nantahala Lake.

"We ought to have had sense not to start that time of day. It was a foolish sort of thing," said Morgan, "but I loved it very much. Everyone was in good spirits."

Sally, Morgan and the NHC often took tools on guided hikes to clear away underbrush. Standing on the old Siler homeplace, Albert Rufus Morgan III said, "I remember Granddaddy riding me on his shoulders when he climbed those mountains, while he swung a machete to clear the trails." In the shadows of the Nantahala Mountains, a well-tended yard surrounded the foundations of "Nonah," built by Morgan's grandparents. The seventy-foot water well, supposedly built by Cherokee Indians, held its ground.

"One of my many chores," said his grandson, "was to spider down to clean it out." At the grassy edge of the wide lawn, one of North Carolina's oldest dogwood trees still clings tightly to the historic site. In the late spring,

North Carolina's state flower blooms in the spring on the dogwood tree. *National Forests of North Carolina Historic Photographs, D.H. Ramsey Library, Special Collections, University of North Carolina–Asheville.*

its long limbs hold white bouquets of the state flower in memory of a man who loved flowers, ferns and trees.

"He could identify the trees in Nantahala's forests," said Rufus III, "even when his vision was poor. He'd finger a leaf in his hand, feeling its shape, edge and texture." Nearing retirement from the Episcopal Church in the early 1950s, Reverend Morgan had moved from Franklin to Nonah, but in October 1975, his historic home burned to the ground. Even his personal journals were lost.

"There were stacks of them in the dining room," said his grandson. "He had written an entry every day since age fourteen. They were all gone." Morgan and friends built a new cabin, Talohi, near the Cartoogechaye Community Center. He died in 1983 at age ninety-six at the Deerfield Episcopal Retirement Center and Nursing Home in Asheville. Respected for his commitments to church and community, "it was his love of the outdoors," reported the *Franklin Press* in 1983, "that prompted the construction of the A. Rufus Morgan Memorial Trail."

Nantahala Hiking Club continues his legacy today. In collaboration with the USFS and the ATC, the group still maintains the A.T. from Bly Gap, Georgia, to the town of Wesser on the Nantahala River. In 1977 near Wesser, when the construction of the new US 64 highway required an A.T. relocation, the group helped reroute the Winding Stair Gap section. With poplar logs and cedar shingles, the group helped replace Wesser Creek Lean-To with the new A. Rufus Morgan Trail Shelter. In 1979, a year after Rufus Morgan retired as the NHC's Trail Manager after thirty-eight years, vandals burned the Wesser Bald Fire Tower. In conjunction with ATC and the USFS, the club helped to build a new observation platform on the original steel frame. In the 1990s, Jack Coriell, NHC's president, updated Rufus Morgan's 1960 description of the A.T. in NNF for the *Guide to the Appalachian Trail in the Southern Appalachians*.

On Siler Bald, the club also helped the USFS and NCWRC restore eighteen grassy acres. Historically, hogs, sheep and cattle have grazed its open meadow. Since the 1970s, the USFS had retarded nature's successional growth of woody plants with prescribed burns every two to three years, but blackberry brambles kept claiming the peak. When slash-and-burn techniques and herbicides were ineffective in promoting grassy growth, the USFS held a commercial timber sale on the summit. With the trees removed, crews root-raked the area and cleared the brush and stumps with a bulldozer. A helicopter seeded Siler Bald with orchard grass and white Dutch clover.

After Hurricane Opal in 1995, NHC was on the A.T. with chain saws before the power company had restored electrical power to area residents. Thirty volunteers worked six hundred hours to remove nearly four hundred trees. The USFS cleared hundreds more of Nantahala's downed trees, repaired roads and bridges and reseeded areas damaged by seventeen major mudslides. One hard-hit area was the east face of Wayah Bald.

Between 1994 and 2007, club volunteers built six new A.T. trail shelters and modified three others to fit their new "Nantahala Shelter" design. A three-sided log shelter, thirteen feet wide and twenty feet long, provided a sleeping platform for eight hikers, along with a covered cooking area. In 2008, they built a special trail shelter less than a mile north of Wayah Bald. After retirement, Larry and Ann McDuff had hiked the A.T. and other long-distance trails. After Ann died in a bicycle accident in 2003, Larry raised $10,500 to have an A.T. trail shelter built in her memory. Sadly, he, too, died in a bicycle accident in 2005 before it could be built. When NHC learned their story, the club proposed a site at Wayah Bald for a memorial shelter, but the USFS lacked the resources. In the *Smoky Mountain News*, Quintin Ellison quoted USFS Brian Browning: "We just don't have the time or manpower to do a project like this. The Nantahala Hiking Club is a huge asset to the district." After four days of preparation, a helicopter flew the bundled materials to the site. ATC paid $5,000 for the delivery. A plaque at the shelter now pays tribute to Larry and Ann McDuff.

In 2001, NHC built Nantahala's first trail shelter privy at Siler Bald. Ten years later, the A. Rufus Morgan Trail Shelter got its own privy. "The work effort in these projects is like a three-legged stool," said NHC volunteer Bill Van Horn. "The ATC helps with the funds; the land manager, which in most cases is the USFS, helps deliver supplies and coordinate the project; and NHC helps provide groundwork, construction and maintenance." More than one hundred members volunteered 8,856 work hours in 2015. Some participated in ATC's threatened and endangered species monitoring project.

The Nantahala Hiking Club helps build and maintain the shelters on the NNF section of the Appalachian Trail. *Nantahala Hiking Club.*

Others shuttled A.T. thru-hikers to Franklin, the first town designated as an "A.T. Community" in 2010. NHC promotes outdoor education in local schools and sponsors teacher workshops, field trips and ATC "Trails to Every Classroom" Programs. At Cartoogechaye and Summit Charter Schools in 2003, it set up a miniature Appalachian Trail. The club also leads numerous public hikes. As of 2015, twenty-three NHC members had completed the Appalachian Trail. Reverend A. Rufus Morgan may rest in peace.

After leaving the A. Rufus Morgan trailhead, a right turn follows NC 1310 and parallels Wayah Creek. Within a few miles, the highway borders Arrowood Glade Picnic Ground and the LBJ Job Corps Center. At LBJ in the 1930s, CCC corpsmen built maintenance shops for their vehicles. At Arrowood Glade, recruits built fish rearing ponds that held nearly sixty thousand trout during peak seasons. After six months, when the trout had grown about seven inches long, the CCC restocked Nantahala's streams. Picnic tables and grills now sit in deserted rearing pools. Rock steps lead up a knoll surrounded by rhododendron to a rustic, log-beam picnic shelter constructed by the corpsmen.

JOYS OF AN APPALACHIA TRAIL JOURNEY

By Marcia Roland, an A.T. section-hiker; former owner of We Be Hiking, a rental outdoor outfitter; and A.T. public speaker, mroland.nb2010@gmail.com.

Living in Franklin, North Carolina, the Nantahala National Forest is my backyard playground. Gentle footpaths, strenuous mountain trails, day-hikes and overnight adventures, as well as the A.T. section that passes through Nantahala, have been my solace. The beauty and challenges of the forest have helped me overcome many personal life struggles.

In 1993, my first experience on the A.T. was a weeklong backpacking trip with my fourteen-year-old son. That experience led me to decide that one day I would hike the whole 2,174 miles. I began planning when, how and with whom I would hike. In 2001, I answered a Texas woman's ad in the *Appalachian Trail*, requesting an A.T. section-hiker companion. The following year, we began our A.T. experience. Each September, we met at Labor Day and hiked for four to five weeks, covering the miles through fourteen states for fourteen consecutive Septembers. We both faced the daily physical rigors of what the A.T. offers, but we also faced our personal challenges more difficult than steep mountains, stormy nights or raging river crossings.

Over the years that we segment-hiked from Georgia to Maine, I battled breast cancer, ended a forty-one-year marriage, learned I had multiple sclerosis and lost my home and material belongings. One thing my experiences on the A.T. taught me, though, is that a positive attitude and being happy can help you handle the major life changes. Our commitment to meet each year no matter what and to hike and push the body and mind to greater things is what helped me through those difficult times. Beth and I found peace, solidarity and confidence in what we had to do to survive in the woods. Our experiences in the woods helped prepare us for the "real" life challenges in our worlds back home. At first, my goal was all about completing the miles, but over fourteen years, I grew spiritually, emotionally and physically. Even though I lost toenails, I will never lose the deep friendship with Beth and the life lessons learned on the trail.

"CCC Veterans honored during Arrowood Glade Ceremony," reported the *Franklin Press* on Monday, October 17, 1983. Two days earlier, on a beautiful fall day, the USFS and the Macon County Historical Society hosted seventy-one CCC corpsmen and their families on the fiftieth anniversary of the CCC program. They celebrated their contributions to conservation. The recruits had planted trees and reseeded road banks. They had built recreational facilities, bridges, dams and the roads at Rainbow Springs and Cowee Bald.

"All of us who have enjoyed the magnificent forests of WNC owe you a debt," said Barbara McRae, president of the historical society, in the ceremony's opening remarks. As a devoted advocate of conservation issues, U.S. Congressman James McClure Clarke delivered the keynote address, echoing her salute. Organizers wanted to not only reunite CCC recruits and rekindle old memories but also help preserve their personal stories. At the gathering, historians collected oral recordings of the men's CCC experiences.

CULLASAJA GORGE

After leaving Arrowood Glade, NC 1310 heads east toward Franklin. Mountain Waters Scenic Byway (US 64/NC 28) goes toward Highlands through Cullasaja Gorge. With its special plant communities, splendid waterfalls and unique geologic features, Cullasaja Gorge in NNF is one of the most outstanding gorges in the state. In 1999, the North Carolina Department of Environment and Natural Resources evaluated Cullasaja Gorge as a potential site for recognition under the NC Natural and Scenic Rivers Act. Passed in 1971, the act aims to protect free-flowing rivers with outstanding features. In March, its report stated, "The Cullasaja River in Macon County is one of many spectacular gorges that cut through the Blue Ridge Mountains. However, its canyon-like profile, with a steep gradient of approximately 180 feet per mile, makes it unique among these rivers." After Cullasaja River flows through Mirror and Sequoyah Lakes near Highlands, it enters NNF below the Sequoyah Dam. For the next seven and a half miles, the river falls 1,300 feet, where centuries of its aquatic forces have chiseled away less resistant rock forms, creating stunning rapids, cascades and waterfalls.

Now listed on the North Carolina Registry of Natural Heritage Areas, Cullasaja Gorge and its misty waterfalls host some of the most biologically diverse spray cliff communities in the Southern Appalachians. Discovered in Cullasaja Gorge in the 1960s, the West Indies dwarf polypody fern is quite rare. As of 1999, the gorge was the only home of a population of this fern outside the West Indies, nine hundred miles away. Other Cullasaja

Gorge specialties include the Appalachian filmy-fern, northern beech-fern and rock clubmoss.

Its natural resources attracted the Cherokees to settle here for centuries. Near their settlement at the confluence of Ellijay Creek and Cullasaja River, called Sugartown (now the Cullasaja Community), they farmed the open valley and fished the river. Some quarried soapstone in the Fishhawk Mountains. In 1776, however, Sugartown was "one of thirty-six Cherokee towns that had fallen to the torch during the American Revolution," wrote Randolph P. Shaffner in *Heart of the Blue Ridge: Highlands, North Carolina.* "General Griffith Rutherford carried out his cruel raids against Indians suspected of siding with the British."

By 1820, white settlers had begun moving in. Silas McDowell lived in a home he built for his mother. He began exploring the region's geology, mineralogy and botany. As a scientific farmer, he raised new varieties of apples. He named one "Cullasaja" and another "Ellijay." When a hard freeze destroyed his six hundred apple trees, he started a vineyard. As an avid writer and historian, he penned many articles and letters urging others to experience Cullasaja's agricultural and scenic wealth. By February 1875, two real estate developers had responded. Traveling from Kansas, Samuel T. Kelsey and Clinton C. Hutchinson established the town of Highlands.

In the 1860s, a gold mine was opened in the area on Gold Mine Creek. On Cullasaja Creek in 1870, farmers plowed up specimens of corundum in the fields of Hiram Crisp. The next year, Colonel Charles W. Jenks opened Corundum Hill Mine, North Carolina's first commercial gem mining operation. Later, miners discovered the first specimen of gem sapphire in its matrix there. Miners have found numerous museum-quality specimens at Corundum Hill, such as oriental emerald, a very rare color for corundum. Rockhounds have also found semiprecious stones, such as garnets, sapphires, rubies and amethysts, in the area.

On Walnut Creek in 1908, the first major logging began in the Cullasaja region and continued until the USFS purchased Cullasaja Gorge. In the 1940s, salvage harvesting removed the stands of American chestnut killed by the blight.

Driving north to south, visitors leave the former valley of the Cherokees and Sugar Town and enter Cullasaja Gorge. At times, only small roadside pull-offs provide parking on the narrow, curvy road. Two lanes hug the steep walls about 250 feet above the gorge. In 1924, the North Carolina Highway Commission thought the nearly perpendicular granite walls were too dangerous and denied a request for the construction of a Cullasaja road between Franklin and Highlands, but district officials voted to proceed. Work began by 1925. Powered by air compressors, wagon drills bored into the rock-solid surfaces. Steam shovels loaded rubble into farm wagons pulled by mules. Temporary roadside camps housed twenty black prisoners who helped build the road. From rope slings lowered from high cliffs, workers dangled and drilled ten-foot dynamite holes. Once they returned to the upper rim, explosives blasted away at the thick roadblock. Local labor also helped. They used timber cut along the roadway to build bridges. They quarried rocks in the gorge near an area now known as Quarry Falls. Without any serious accidents, crews completed the road in 1929. In the 1930s, the CCC improved the route.

About eight miles east of Franklin, Lower Cullasaja Falls is a dramatic introduction to the beauties that lie ahead. The river exits a steep canyon to dive two hundred feet over a quarter of a mile in a series of powerful cascades. Each cascading stream, with its own distinctive character, races

Crews use a steam shovel at the quarry in Cullasaja Gorge to build the highway between Franklin and Highlands. *George Masa Collection, Highlands Historical Society.*

headlong, leaping ledge to ledge, to join as one in the gorge below. For a premier performance, visit Lower Cullasaja Falls after a heavy rain or in the autumn, when surrounding hardwoods cloak the falls in a kaleidoscope of color.

A two-mile drive upstream leads to Quarry Falls. At the site of the former quarry used to build the highway, there is a wide parking area. With the help of the USFS, an Eagle Boy Scout with Troop 202 in Franklin provided an information board for maps and exhibits. A boulder-strewn section of Cullasaja River draws sunbathers and picnickers in the summer. Locals call the twenty-foot, three-tiered waterfall "Bust-Your-Butt Falls." Although this area provides access to the river's edge, sound judgment on slippery surfaces is necessary.

Within a few miles, USFS Cliffside Lake Recreation Area offers a picnic shelter built by the CCC. In its classic style, heavy log beams lift the rooftop above rock walls and a chimney. Years ago, USFS assistant ranger Bruce Bayle led a project to fill in the four original CCC grills with local stone. Visitors make reservations for group shelters through the Cradle of Forestry in America Interpretive Association. Numerous open picnic tables are freely available. Beside one table, a lone chimney remains. According to Bayle, the CCC used cribbing construction to build it next to a privy. To remove accumulated waste material, custodians added diesel fuel and struck a match. Disintegrated waste wafted out the chimney.

A CCC-constructed dam across Skitty Creek forms a six-acre cold-water lake for swimming and fishing. The NCWRC annually restocks the lake with trout. From the parking lot to the lake, a wide PVC pipe sends trout down a safe chute to simplify the task. Near the lake's sandy beach, the CCC built a bathhouse and a large wooden swing set. An Eagle Scout with Franklin's Troop 229 helped the USFS restore it in 2004. Six short trails are available at Cliffside, including a family-friendly loop around the lake. Some are handicapped-accessible. Clifftop Vista Trail climbs one and a half miles to a CCC-constructed wooden gazebo. The easy half-mile Potts Memorial Trail leaves the Vista Trail through a white pine plantation. A half-mile pathway connects Cliffside to the USFS Van Hook Glade Campground.

Near Cliffsides on the scenic byway, a large parking area provides access to Dry Falls. Originally known as "Cullasaja Falls," the CCC renamed it. While building rock walls for a pedestrian walkway, corpsmen noticed the rock ledge launching Cullasaja seventy-five feet into a deep, thunderous chasm. Over the centuries, turbulent water had scooped out less resistant rock types beneath the shelf of quartz biotite gneiss. The alcove that formed behind the

Right: Dry Falls before the CCC built the visitor's walkway behind it. *National Forests of North Carolina Historic Photographs, D.H. Ramsey Library, Special Collections, University of North Carolina–Asheville.*

Below: The CCC built rock retaining walls, observation platforms and a walkway behind Dry Falls in the 1930s. *National Forests of North Carolina Historic Photographs, D.H. Ramsey Library, Special Collections, University of North Carolina–Asheville.*

Years ago, the highway between Franklin and Highlands used to curve behind Bridal Veil Falls. *National Forests of North Carolina Historic Photographs, D.H. Ramsey Library, Special Collections, University of North Carolina–Asheville.*

falls kept visitors dry. Looking through the waterfall from the open chamber, visitors get a unique river-scape perspective into the gorge from Cullasaja's point of view. During the restoration of retaining walls, safety barriers and viewing areas, the USFS has replicated CCC craftsmanship.

Beyond Dry Falls, the highway passes McCall Cabin and approaches Bridal Veil Falls. Motorists used to curve behind the 120-foot slender waterfall. Now the wide parking area and semicircular pavement behind the falls permit a stroll to scan high banks and sprayed cliffs for special plants. Icicles glisten in winter. The stream of water flows under US 64 to join Cullasaja in the gorge. Beyond the Bridal Veil, a series of small falls drops for a quarter of a mile below Sequoyah Dam near Highlands.

Two miles west of Highlands, the town built Sequoyah Dam and powerhouse in 1927 to provide the town's water and electricity. When the powerhouse proved too expensive to maintain, it was closed in 1968. When workers drained the lake in 2015 for dam repairs, they found an eighty-five-year-old wooden boat, half buried by sand and clay. In the 1930s, when

Kenyon Zahner hired local craftsmen Joe Webb and Furman Vinson to build a cabin on the new lake, they designed a boat landing underneath the home. Builders used this boat to ferry out dirt and sand. Overloaded on its last voyage, it sank. Highlands Historical Society rescued the relic, which is displayed at its historic village.

Before exploring Highlands, consider a side trip on Buck Creek Road. A strenuous nine-mile hike leads to a fire tower, listed on the National Register of Lookout Towers, atop the 5,127-foot Yellow Mountain. In 1934, the USFS and CCC built its seven-and-a-half-foot rock base and wooden cab. In the 1980s, the USFS restored the tower, now dedicated to Pat McClure, a fire warden for nearly ten years, until it closed in 1969. An observation platform has replaced the wooden cab. Also in the 1980s, wildlife officials reintroduced peregrine falcons here. Nearby, the North Carolina Forest Service acquired a USFS special-use permit for seed reproduction of Fraser firs on three to five hundred acres.

HIGHLANDS-CASHIERS

"Are you doing the circuit?" asked a retired nuclear engineer from Charleston, South Carolina, pausing for a conversation on the Glen Falls trail. "This is our last day here in the mountains. We're circling the waterfall tour. After this one, we're headed to Cullasaja Gorge."

In May 2016, families wandered down the two-mile roundtrip trail outside Highlands to see the spectacular views of Glen Falls. Shaded by white pines and rhododendron, the wide trail led to wooden observation decks built by the USFS. Recently back from the tropics, a wood thrush flitted limb to limb in the dense understory, escaping a binocular view. A downy woodpecker drummed a dead hemlock. Red-breasted nuthatches played among the pines, unaware that it was Mother's Day. Water and rock, light and sound created a dazzling show of natural beauty from several platforms along the long, steep descent of the East Fork of Overflow Creek. The Glen Falls Trail was a great start to a day exploring NNF near Highlands. A drive northwest of Highlands tours the waterfalls of Cullasaja Gorge, and a drive southeast descends the Highlands plateau into peaceful Horse Cove.

In an 1844 North Carolina land grant, Captain Joseph W. Dobson of Cartoogechaye acquired two tracts of land, each with 640 acres, on the plateau now known as Highlands. He built a cabin and hired a couple to oversee his grazing livestock. Thirty years later, this area attracted the interest of the two developers from Kansas. Did Samuel Kelsey and Clinton Hutchinson draw lines on a map, as tradition claims, from Chicago to Savannah and from New York City to New Orleans to find the intersecting, geographical center of eastern America? Were they attracted to the thermal belt growing conditions advertised by Silas McDowell of Cullasaja? No doubt the men believed in the curative benefits of cool, high-elevation mountain air. After creating the town of Hutchinson in flatland Kansas, the two men rode mules to the top of the 4,118-foot plateau. In 1875, their pocket compass, now on display at the Highlands Historical Society Museum, determined the boundaries of their original 800 acres purchased from Dobson.

Kelsey promoted their new town in pamphlets distributed across the country. New residents bought land and stayed. As families settled in Highlands, Hutchinson moved back to Kansas, and Kelsey surveyed land near Grandfather Mountain to establish Linville, North Carolina. Farmers, gold miners, cheesemakers, beekeepers and businessmen continued to migrate to Highlands. Reverend Potts moved from Franklin with one hundred horses to start a livery stable. Dr. Mary E. Lapham treated tuberculosis patients in the sixty cottages of her Highlands sanatorium by compressing one lung. Scientists and explorers also came. T.G. Harbison came south on a botanical expedition. Walking from Pennsylvania to explore the Southern Appalachians in 1886, he became a civic leader and schoolteacher in Highlands. When officials planned to close the school for lack of funds, Harbison led his students on botanical collecting trips. They earned money for the school by selling plants to northern markets. By 1896, Harbison was collecting plant specimens for George Vanderbilt's Biltmore Estate.

Private sawmills met the growing need for lumber. Innkeepers built boardinghouses to accommodate visitors. A letter sent home from Horse Cove by Woodrow Wilson, as quoted in a local newspaper at the time, confirmed the twenty-three-year-old's stay at the Thompson Inn in 1879. As more people moved to Highlands, it became the highest incorporated town in the East. NNF became the resource manager of nearly 60 percent of its surrounding forests.

Southeast of town, the main street in Highlands continues past the Highlands Biological Station, opened in 1927 as a museum to preserve Indian artifacts, minerals, plant specimens and other natural history collections of early residents. Over the decades, the station has become an environmental research center for scientists from around the world. The adjoining chestnut and granite museum now houses natural history exhibits, including a 425-year-old slab of a hemlock tree. The center offers public lectures, guided hikes and children's events. Beside a lake, a five-acre botanical garden contains hundreds of plants in a variety of habitats.

An early view of Main Street in Highlands. *Macon County Historical Society and Museum.*

Across from the Highlands Biological Station, a short trail leads to Sunset Rock in Ravenel Park, a high rock face given to the city by the Ravenel family in 1914. In 1879, the Ravenels of Charleston built the first summer home in Highlands. At Sunset Rock, a western view overlooks Highlands. A faint trail to the east side of the rocks provides a sunrise view of Horse Cove and NNF.

Continuing southeast, Main Street soon becomes Horse Cove Road as it leaves the city limits. For the next two miles, the roadway twists through more than thirty hairpin turns as it descends nearly one thousand feet. A brief detour on Rich Gap Road (FSR 401) between Fodderstack and Rich Mountains leads to a short trail to a living memorial for USFS Ranger James R. "Bob" Padgett. The plaque at the base of the poplar tree states that Padgett "in 1966 saved this magnificent tree…with a girth of twenty feet, a height of one hundred and twenty-seven feet and an age of four centuries."

At the parking area, a large exhibit celebrates the work of the CCC. More than two hundred recruits transferred from an encampment at Dahlonega, Georgia, to establish the Horse Cove CCC Camp F-19. Recruits helped with USFS projects at Dry Falls, Ammons Campground, Cliffside Lake and Van Hook Campground. The men upgraded Cullasaja Gorge Road and transplanted honeysuckle vines to reduce erosion. In Horse Cove, they quarried stone to build Bull Pen, Horse Cove and Whiteside Cove Roads. After scientific studies at Coweeta Hydrologic Laboratory determined better road designs that reduced maintenance and helped water runoff, explained USFS Chad Boniface, the USFS improved CCC-constructed roadbeds using the "Coweeta Grade," with dips and curves to improve drainage and prevent washboard surfaces.

Farther along Horse Cove Road, the granite walls of Blackrock Mountain, which once supported a circa 1900 lookout tower, rise above the green valley. Bull Pen Road veers right toward the Slick Rock Trail overlook, USFS Ammons Branch Primitive Campground, Ellicott Rock Wilderness and Chattooga River. A left turn follows Whiteside Cove Road, where the fun at heart climb a short but steep USFS trail to the fissures, boulders and caves of Granite City at the base of Blackrock Mountain. East of Whiteside Cove Road, but hidden from sight and sound, the Chattooga National Wild and Scenic River courses through narrow channels beneath high cliffs. Soon, the road crosses Norton Mill Creek, seeping to the surface at the base of Whiteside Mountain and flowing south to join the Chattooga.

Whiteside Mountain. In 1929, when George Masa was taking photographs for a *National Geographic* article by Horace Kephart, Frank Cook, the owner of Highlands Inn, hired him to photograph scenes around the Highlands Plateau to inspire tourists to visit the area. Those one hundred photographs are displayed in the Frank Cook Gallery of the Highlands Historical Society Museum—the largest collection of his historic work that remains today. *Highlands Historical Society, George Masa Collection.*

After rounding a curve, motorists are greeted with one of the grandest sights in eastern America. The rock walls of Whiteside Mountain rising more than 1,800 feet above the valley floor reflect in a large private lake. A fall color experience is unforgettable. In 1827, when the Cherokees occupied the area, Barak Norton, the first white settler in the cove, built his log cabin at the base of Whiteside. He named his daughter Sarah Whiteside Norton, the first white child born here. By 1880, North Carolina had awarded at least twenty land grants to pioneers settling around Whiteside Mountain.

Considered one of the oldest mountains in the world, Whiteside Mountain (4,930 feet) is a pluton, a geological term derived from Pluto, the Roman god of the underworld. Millions of years ago, giant tectonic plates of the earth's surface pressed together; some landforms rose into mountains, while others cracked apart into deep fissures. Deeper cracks into the earth's interior caused molten material to seep into the crevice. Slowly, underground molten rocks cooled and hardened into granite. Extreme heat and pressure changed some granite rocks into gneiss. Eons of earth-altering events and erosion exposed the domed walls of the pluton seen today.

In the late 1800s, Margaretta, wife of Captain Samuel Prioleau Ravenel Sr., and her son, Prioleau Jr., began buying tracts of land on and around Whiteside Mountain from early settlers. They purchased thousands of surrounding acres from Macon Lumber Company. In 1869, the logging company had received the property in a thirty-thousand-acre state land grant. In 1902–3, Prioleau created a toll road, improving the access from Cashiers to Highlands and allowing the public to continue visiting the family's mountain. At Margaretta's death in 1913, her son inherited the estate.

In about 1914, when the USFS began buying land to protect valuable watersheds in the new NNF, the government paid $7.50 per acre for more than 700 acres at the base of Whiteside Mountain. The USFS wanted to protect the headwaters of Norton Mill Creek and Chattooga River as well. Another area of interest was the old Kelsey Trail, which passed through a 1,500-acre old-growth forest, known locally as the "Primeval Forest." In the 1930s, the Highlands Biological Station and the North Carolina Academy of Science urged the government to purchase this biological treasure. Prioleau, however, chose not to sell it.

When Prioleau died in 1940, his second wife, Beatrice, inherited Whiteside Mountain. With little interest in the mountain and its surrounding natural beauty, she sold more than 1,500 acres in 1943 to Powell Lumber Company and Champion Paper and Fibre Company. Scientists with the Highlands Biological Station and a USFS botanist bitterly opposed the deal, but it was too late. By 1946, acres of birch, hemlock, cherry and maple on the western slopes of Whiteside Mountain were gone. "The Upper Cullasaja River ran brown with silt," wrote Robert Zahner, author of *The Mountain at the End of the Trail: A History of Whiteside Mountain.* "Hundreds of thousands of forest interior birds and animals lost their habitats—their homes—as the big trees were felled, bucked into logs or split into pulpwood and hauled away to Waynesville and Canton."

After the companies harvested many of Whiteside's big trees and were ready to sell in 1948, Congress could not provide the funds to purchase it. Whiteside Mountain Inc. bought the mountain and transformed it into a tourist attraction. During the 1950s, a one-dollar toll fee allowed road access to a parking lot and concession stand. Twenty-five cents bought a ticket for a jeep-led tram ride to a wooden platform at Whiteside's summit. In 1953, a forest fire burned the slopes of Whiteside Mountain for four days, and by the 1960s, the tourist business had declined. Whiteside Inc. asked $300,000 for its mountain. The U.S. government was interested but again lacked the funds. Congressman Roy Taylor and NNF supervisor Del Thorsen received an approval from the chief of the Forest Service for a land exchange deal. In the 1970s, Andrew McDonough, a Highlands real estate developer, bought Whiteside Mountain and exchanged it for USFS land located adjacent to his property on US 64.

To protect the headwaters of Chattooga River, recently designated as a National Wild and Scenic River, the USFS expressed interest in the eight hundred acres on Devils Courthouse. Inherited by Dr. George E. Crouch IV in 1969, the owner agreed to sell the tract to the USFS with one stipulation. At the overlook along US 64, known as the "Big View," he requested that the USFS place a plaque in memory of his father. Congress approved the funds, but the state could not grant permission for the roadside plaque. When Crouch died fifteen years later, his widow sold the property to the USFS for six times the previous price. In memory of her husband and father-in-law, the USFS agreed to place a plaque at an overlook on Whiteside Mountain.

If a lifetime only permits the climb of one mountain, this is it. After the USFS acquired Whiteside Mountain in 1974, the agency, with the help of the Youth Conservation Corps, built a new trail to the summit. Visitors access the two-mile loop at the USFS fee area on Whiteside Mountain Road near US 64. Now a National Recreation Trail, the five-hundred-foot climb follows the old jeep/tram route. It passes an old dividing spring near a former campground that previously sent part of its waters south to Chattooga River and the Atlantic. The rest went north to the Cullasaja

River and the Gulf of Mexico. When the USFS built its new visitor parking lot, it diverted the double-spring waters toward the Chattooga. Naturalists find a botanical paradise within Whiteside's fourteen natural communities. Specialties include Hartweg's locust, deerhair bulrush and various sedges. There are speckled wood lilies, Buckley's St. John's-wort and Curtis's goldenrods. Strong winds and winter storms have reshaped and twisted old-growth northern red oaks. The heath communities contain minniebush, sand myrtle and pinkshell azaleas.

The granite domes of NNF's Terrapin Mountain and the privately owned Chimneytops and Rocky Mountain are seen from the open cliffs. A short but challenging spur trail leads to Devils Courthouse. Peregrine falcons, reintroduced in the 1980s in a collaborative project between the USFS and the NCWRC, have reestablished aeries here over the decades and can be seen gliding along the cliffs.

"Last year, I saw a bear down there!" exclaimed a ten-year-old boy, pointing over the railing beside the trail.

"We were looking over the valley," said his father, "when we saw a black head rise above that steep dropoff. Apparently, he was standing on a ledge down there."

Beside the trail at Fool's Rock, a USFS exhibit describes a fall over Whiteside's sheer rock wall in May 1911. Gus Baty landed 160 feet below on a narrow ledge. Charles Wright climbed down to save him. The U.S. government presented him with the Andrew Carnegie Gold Medal for Bravery. From the summit, the Whiteside Mountain Trail loops back through a beautiful forest and returns by switchbacks to the parking lot.

On the Whiteside Cove Road, motorists pass Grimshawe's Post Office, which served the community for fifty years. Under a British flag, an Englishman opened the post office in 1903. As a tourist attraction, Whiteside Mountain Inc. moved it up to the Whiteside Mountain parking lot in 1953. After the USFS purchased the mountain, the agency moved it back. A right turn onto NC 107, south of Cashiers, leads past a USFS trailhead to thirty-foot-high Silver Run Falls on a tributary of Whitewater River. The headwaters of Whitewater River rise on High Hampton Inn's Chimneytop Mountain.

Shortly after NC 107 crosses the South Carolina border, a sign directs visitors to a road leading back into North Carolina to the spectacular Whitewater Falls, the East's highest waterfall. In 1974, the USFS, Duke Power Company, the Foothills Trail Conference and others created the seventy-seven-mile Foothills Trail that passes Whitewater Falls. A 150-step wooden staircase on the Foothills Trail leads to a closer view. After three attempts, a Blackhawk helicopter successfully lowered an iron bridge on the trail near the base of the falls to continue the Foothills Trail across Whitewater River. At the USFS picnic area, a handicapped-accessible trail leads to a waterfall overlook. Scientists have studied the botanical diversity of mosses of Whitewater River Gorge. "Its 268 species of mosses," wrote Tim Spira in *Waterfalls and Wildflowers in the Southern Appalachians*, "represent about one-quarter of the total known species of mosses in eastern North America."

CHATTOOGA NATIONAL WILD AND SCENIC RIVER

South of Cashiers on Whiteside Cove Road, retired USFS Chad Boniface narrated a trip into the Chattooga region. At a small pull-off, four miles south of Cashiers, the first glimpse of Chattooga's headwaters rushed over "Nantahala's Sliding Rock." On that July day, Chattooga's shallow flow pushed young boys down a smooth bank-to-bank rock face into a swimming hole. Upstream, Chattooga had passed through a secluded pocket in NNF; downstream, it would wind through private property on its journey to Ellicott Rock Wilderness, the nation's only federally protected wilderness shared by three states. In North Carolina's section of the wilderness, the 3.5-mile Bad Creek Trail descends to the Chattooga River on steep switchbacks. The 4.3-mile Ellicott Rock Trail descends gradually on an old roadbed, crosses into Georgia and requires a tricky ford across Chattooga to reach the historic Ellicott Rock.

Farther south, Whiteside Cove Road ends at Horse Cove Road. After turning left onto Horse Cove Road, FSR 1178 veers right and crosses the Bull Pen Bridge spanning the Chattooga River. A bridge-side view offers an overhead perspective of nature's power, sculpting potholes and sanding large boulders. The bridge hides a small waterfall foaming underneath the rustic iron overpass.

The Chattooga Cliffs Trailhead is located nearby. Overlapping canopies of oak, maple, beech and hemlock trees shade the dense understory of rhododendron, dog hobble and mountain laurel. Wildflowers flourish in spring and summer. About half a mile from the trailhead, the two-mile loop built by the Youth Conservation Corps leaves the river, climbs the ridge and returns to Bull Pen Road. An alternate choice continues straight ahead. This five-mile trek roughly follows the banks of Chattooga River. It then climbs the gorge on an old roadbed and exits at a parking area near a church on Whiteside Cove Road. Although vegetation and large boulders occasionally limit views of Chattooga River, its roaring din often penetrates the forest like a thunderous applause. Anglers and waders gain access at bridge crossings, giant rock slabs, deep pools and sandy beaches.

At times, slippery terrain and rocky overhangs slow a hiker's pace. Some rough areas require roothold scrambles. In the 1990s, the USFS eliminated precarious rock-hopping sections across two of Chattooga's tributaries. A helicopter lowered iron bridges along the trail at the mouth of Cane and Norton Mill Creeks. Scenic cascades, deep pools and tabletop rock slabs at Norton Mill Creek tempt an unplanned detour.

Farther north, Chattooga's bolder nature shoots through a fifty-foot-long, six-foot-wide raceway. Beyond a cascading tributary, the Chattooga Cliffs border the river for nearly a mile. The Chattooga Cliffs, and an area known as the Narrows, support specialties like the divided-leaf ragwort and Appalachian clubmoss. Filmy ferns, more common in the tropics, also grow in the Chattooga watershed. Colonies of flame azaleas brighten the understory. Along vertical rock walls sprayed by waterfalls, rare mosses, liverworts, algae and vascular herbs cling to shallow soils. L.L. Gaddy stated in *The Naturalist's Guide to the Southern Blue Ridge Front* that "from a botanical point of view, this area is one of the most noteworthy on the Chattooga."

Beyond Bull Pen Bridge on the Forest Service road, the Chattooga River flows south, tracing the South Carolina–Georgia state line. For decades, paddlers have navigated the Chattooga River's lower section to its mouth at Lake Tugalo. When filmmakers brought James Dickey's novel to life in 1972, thrilling audiences with its whitewater adventures, the movie *Deliverance* attracted more people to the Chattooga River. Commercial operators began offering more guided rafting trips. Some paddlers, though, began campaigning for access to its upper gorge in NNF.

After Congress designated a portion of the Chattooga as a National Wild and Scenic River in 1974, protecting it as a free-flowing river in its natural state, NNF managed Upper Chattooga as a recreation area for hikers and anglers. However, federal protection under this act did not prohibit boater access. After years of debate, NNF consented to limited boating access to the secluded upper channel. American Whitewater Association contested the USFS restrictions and filed a lawsuit.

In 2011, the USFS conducted an environmental assessment of the Upper Chattooga to evaluate the potential recreational boating impact on the region's threatened and endangered species. For thirty-five years, researchers had studied the watershed's ecosystems, unique plant communities, soil characteristics and spray cliff habitats. After evaluating data from past research and current field analyses, the USFS determined that watercraft activities in the Upper Chattooga would not affect any identified protected species.

In 2012, the court's decision directed the USFS to regulate paddler access to seventeen miles of the headwaters region. During daylight hours between December and April, when water flow rates are within acceptable limits, paddlers can now navigate Upper Chattooga River. Groups are limited to six boaters, and all individuals must register at designated launch sites.

BALSAM MOUNTAINS

In 1981, the USFS purchased the forty-thousand-acre Balsam–Bonas Defeat tract. At the time, Lewis Kearney was working at the USFS Southern Regional Land Acquisitions Office, reviewing land appraisals and allocating funds to southeastern national forests. When officials finalized the purchase, the Government Printing Office requested that someone collect the large check in person. Kearney flew to Birmingham, Alabama, returned to Atlanta and casually strolled into the office.

"Where's your briefcase with the check?" asked his boss.

"Oh that," replied Kearney, fumbling through pockets of chewing gum, reading glasses and crumpled papers. "It's in here somewhere....Oh, here it is," he said, handing the $13 million check written to Carolina Riteco to the worried boss.

The purchase included the Bonas Defeat acres owned by James McClure Clarke, as well as the Balsam tract formerly owned by the Mead Paper Company. According to Bill Leatherwood, son of a former Mead Company supervisor, Armour Leather Company owned a tannery in Sylva in the 1920s. Chestnut trees harvested at Balsam provided tannic acid for processing its leather products. Cattle raised near Sylva provided the hides.

"From Scott's Creek to the Tuckasegee River," said Leatherwood, "the water would run black from the tannic acid....The Armour Company wanted to find a use for the leftover wood chips after the tannic acid was removed.... That's when Mead Paper Company came and the two companies worked together until the tannery left." Mead operated a paper mill in Sylva for thirty-seven years.

Mead had purchased the Balsam timberland from Blackwood Lumber Company. The Blackwood Company logged most of the large poplar and chestnut trees for its lumber mill near East Laporte between Cashiers and Cullowhee. Joe Keys, the mill's owner, died around 1940, explained Gerald Ledford, a local logging and train historian. Keys's widow kept the Blackwood mill and railroad running until the end of World War II. When Blackwood closed its operations, the company sold its property to the Mead Paper Company.

In the Balsam Mountains, Mead built a recreational retreat for its executives. Overlooking an eight-acre mountain lake, created by damming Wolf Creek, the company constructed its five-bedroom Balsam Lake Lodge. In the living room, a huge rock fireplace faced the opposite wall of floor-to-ceiling windows. Chestnut walls, oak floors and a pine ceiling graced the large dining room. Each bedroom had a name, representing the wood used to panel its interior walls: the Ash Room, Cherry Room, Chestnut Room and Buckeye Room. Cedar wood lined each bedroom closet. As a supervisor for the Mead Company, Leatherwood's father brought the family here for vacations in the 1950s. Years later, Carolina Riteco acquired the Balsam property.

When the USFS purchased the Carolina Riteco property, the tract included the lake and lodge. The agency built a picnic area and hiking trails. Near the picnic area, a former superintendent's cabin became a seasonal home for USFS recreational hosts. Since 2008, Sylvia and Don Snook have volunteered at Balsam Lake, living in the chestnut-lined cabin for one to two months a year.

Wolf Creek flows beside a picnic shelter and enters Balsam Lake under an arched bridge. Beneath a clear October blue sky, the mirrored image of fall-colored hardwoods doubles the lakeside charm. The USFS built handicapped-accessible trails, fishing piers and platforms, as well as a wheelchair-accessible shower room in Balsam Lodge.

"During the remodeling phase, the Forest Service placed a high priority on making Balsam Lake available to the disabled," said USFS Chad Boniface. "It is the most handicapped-accessible recreational facility in NNF."

On a special Kid's Fishing Derby Day, USFS staff lowered weighted nets near the arched bridge to corral fish in the stocked creek. Floats on the other end of the nets kept them in place. Children with disabilities dropped fishing lines into the pools of fish to hook a big surprise.

"The huge smiles on those kids' faces," said Boniface, "was enough to keep your life glowing for months."

Sunrays over the Great Balsams. *Kristina Plaas, Plaasabilities Photography.*

In 2013, *Smoky Mountain News* reported, "Balsam Lake is high and dry as the tourist season hits full stride." Over the winter months, a stress fracture in the dam had drained the lake, the second breach in two years. After a $15,000 investment and long delays from heavy rains, contractors repaired and reinforced the dam's splashboard. Wolf Creek refilled the lake and the NCWRC restocked it with fish.

NNF now includes the Balsam property as part of its Roy Taylor Forest. The Blue Ridge Parkway borders the Roy Taylor Forest to the north and the Pisgah National Forest to the east. The Mountains-to-Sea Trail, which will stretch from the Smokies to North Carolina's coast, parallels the parkway and makes its sole appearance in NNF as it crosses the five-thousand-foot highlands.

PANTHERTOWN VALLEY/BONAS DEFEAT

In 1966, after thirty years of construction, road crews had completed nearly all of the 469-mile Blue Ridge Parkway from Shenandoah National Park to the Great Smoky Mountains National Park. Controversies over the 7.7-mile stretch around Grandfather Mountain delayed the official dedication for twenty-one more years. The anticipation of completing the long-awaited scenic parkway helped spark a broader idea. In 1961, Congressman Roy Taylor had proposed adding an extension to the Blue Ridge Parkway. After two years of surveys and negotiations, politicians and local officials met in Highlands on December 11, 1963, to discuss the 190-mile extension from the Blue Ridge Parkway at Beech Gap to Cartersville, Georgia. The planned route would pass Whiteside Mountain, Bridal Veil Falls, Cullasaja Gorge and Estatoah Falls and cross the Chattahoochee River in Georgia to end at a site near Atlanta. In Jackson County, North Carolina, the Georgia Parkway Extension would pass through the valley of Panthertown. By January 1968, the federal government had passed a bill authorizing $87.5 million to build the southern extension. By October, President Lyndon Johnson had signed it into law.

In 1970, the owner of Panthertown Valley, Liberty Properties Corporation, a subsidiary of the Liberty Corporation, did not object to the nine-mile section of the extension coming through its property. It just opposed the designed route that would slice through "the heart of our planned development of second-homes, lakes, golf courses, etc.," wrote David W. Robinson II, Liberty's legal vice-president, in a letter to Senator Ernest Hollings.

Not only did Blue Ridge Parkway planners hope to bring the road extension through the central valley, but they also wanted to build a large campground in Panthertown. Liberty offered to "cooperate fully" with right-of-way passage for the parkway across its southernmost ridge and allow the parkway to build its campground there. Tourists traveling on that section of the proposed Blue Ridge Parkway would have great views into South Carolina and a convenient access to Liberty's proposed lodge, golf course and lake in the valley. With Liberty's experience in operating Lake Arrowhead Campground at Myrtle Beach, it wanted the parkway to build its campground at Panthertown's southern edge and allow Liberty to manage it.

Congress required parkway officials to acquire the 43 North Carolina miles and the 137 Georgia miles of the extension by donation only. About 82 uncontested miles traversed federally owned land, mostly managed by the USFS. The remaining 100 miles of right-of-way required the landowners' support. City officials in North Carolina and Georgia supported the project for its potential increase in tourism, but environmentalists opposed it. The Appalachian Trail Conservancy objected to its impact on the trail in Georgia. Corporations, like Liberty, insisted on more money for the right-of-ways than the U.S. government could afford. Many private second-home owners avidly protested its construction.

In 1978, Western Carolina University professor Dan Pittillo was studying rock outcrop vegetation in Panthertown during the controversy. He and the Blue Ridge Parkway superintendent stood on a cliff overlooking the valley and its high rock domes and granite cliffs and discussed the importance of protecting this beautiful area and its valuable ecosystems as public lands. However, at a public meeting at Cashiers, the National Park Service, the managing agency of the Blue Ridge Parkway, rolled out a map showing the route of the planned extension.

"When I and others saw that a proposed lake would flood Schoolhouse Falls, we objected strongly," said Dan. The development plans would destroy some of Panthertown's most significant features. As time passed with mounting contention and debate, government funds dwindled. Two years before parkway officials settled the Grandfather Mountain issues and officially opened the original Blue Ridge Parkway in 1987, organizers abolished the idea for a parkway extension to Georgia.

In the 6,300-acre Panthertown Valley, Liberty Properties never built its dam across the headwaters of the Tuckasegee River, its mountain lodge on an open rock cliff or its golf course in a flat, sandy valley. It never cut the

Schoolhouse Falls is one of the most visited sites in Panthertown Valley. *Chris Berrier.*

hardwood forests recovering from previous logging operations, fires and floods. Its development never destroyed rare mountain bogs and a diverse range of plant communities. Uncommon species, like Cuthbert's turtlehead and rock gnome lichens, continued to thrive. So did rare mosses and ferns in the misty sprays of nearly a dozen gorgeous waterfalls.

Soon after the parkway extension plan fizzled, the Liberty Corporation decided to sell Panthertown Valley. The USFS and the Nature Conservancy expressed interest and negotiated its purchase. So did Duke Energy, which wanted to build a power transmission line through the valley. When Liberty accepted Duke Energy's offer, environmentalists filed a lawsuit in an effort to stop the sale but lost the case in an Asheville courtroom. With little delay, the North Carolina chapter of the Nature Conservancy approached Duke, raised the funds and purchased Panthertown Valley from the power company with one stipulation: Duke wanted to construct its transmission line through the valley and retain its corridor, but it agreed to build the power line up high and paint the towers in neutral colors. After Congressman James McClure Clarke, a staunch supporter of Panthertown's protection, acquired $8 million from the Land and Water Conservation Fund within the next year, the USFS bought Panthertown Valley from the Nature Conservancy to become part of Nantahala National Forest.

NNF now manages Panthertown as a backcountry area. Except in a few limited situations, the Forest Service has halted timber harvesting there. It prohibits motorized vehicles on the trail systems. In a partnership with the Friends of Panthertown Valley, the Nantahala Ranger District has mapped and signed about thirty miles of official USFS trails. They have placed maps and exhibits at major access points at Cold Mountain Gap and Salt Rock Gap. Many wide trails follow old logging roads. The Powerline Road trail welcomes hikers, bikers and equestrians. Mac Gap Trail passes through a white pine plantation planted in the 1960s by Liberty Properties. Below Big Green Mountain, a trail provides views of the Great Wall, a one-mile-long, three-hundred-foot-high granite cliff. On the Great Wall, rock climbers get a breathtaking experience that others miss. Some trails lead to distant views from open cliffs. Others border lazy tea-colored creeks, high-elevation bogs, white sandy beaches and stunning waterfalls. A convenient access to Schoolhouse Falls makes it one of the most popular destinations. Panthertown, Greenland and Flat Creeks provide twenty miles of catch-and-release fishing for native brook trout. The Panthertown Valley Trail passes the confluence of Panthertown and Greenland Creeks, the beginning of Tuckasegee River. Primitive camping is available nearby.

Experienced backcountry hikers explore some of Panthertown Valley's unofficial trails. In the northern section, unmaintained trails lead to Panthertown's highest waterfall, the 250-foot-high Flat Creek Falls. Some of the obscure trails still trace footpaths cleared by Carlton McNeill. An unassuming, well-intentioned local resident who loved Panthertown Valley, McNeill offered himself as a free guide in the 1990s to anyone who wanted to see his huge, "adopted" backyard valley near his cabin at Cold Mountain Gap.

On a crisp, sunny October day in 1996, Carlton met a family of six from Asheville at his home. Across from the entrance to Canaan Land Christian Retreat on Cold Mountain Road, he lived within walking distance of Panthertown Valley. As the group approached, he introduced himself as the "unofficial guide of Panthertown" and handed them his hand-drawn trail map and business card.

"You got everything you need?" he asked. "We'll be gone all day."

Turning to the younger ones, he said with a twinkle in his eyes, "Now, it's real rugged in Panthertown; you gotta' be *mean* to go in the valley!" A small, elderly man with a big smile and winning personality, Carlton's love for Panthertown soon became contagious. With a tall stick and a fast pace, he led naïve hikers down ravines, up steep rock domes and past beautiful waterfalls, patiently waiting for the less hardy to catch up. On a whim, he would push through thick rhododendron and laurel bushes to take shortcuts. "I said a shortcut," he chuckled, "not an easy cut!"

He not only built trails but also observed others who spent time in his valley. He spoke of Dan Pittillo studying the rock outcrop vegetation, grottoes, wetlands and seepage areas; Allen Boynton with the NCWRC, monitoring peregrine falcons; and groups of botanists cataloguing Panthertown's wildflowers. On that day's hike, Carlton stopped to converse with twenty students from Nantahala High School.

According to USFS Chad Boniface, the Nantahala Ranger District began an inventory of potential trails for Panthertown Valley in 1996. Although Carlton's hand-cut trails had provided access to scenic wonders, some of his trails passed through fragile ecosystems. Trails that he had cleared straight up a mountainside caused runoff and erosion. Others were unsafe for public use.

"Carlton's trails created some challenges for us," said Chad. "The Forest Service had to provide the public with a safe network of trails that could be maintained. Some of his trails followed old logging roads, but others had to be rerouted to bypass protected areas." One of his trails climbed Big Green Mountain too close to a rare granitic-dome plant community.

The Nantahala Ranger Station provides official USFS maps of the blazed trails in Panthertown. Burt Kornegay, a Panthertown expedition guide for more than twenty years, created a map that includes the fainter, unmarked trails for experienced hikers in the wilderness areas. A Global Positioning System (GPS) and compass are also recommended.

In the Valley of Panthertown, the Tuckasegee River gathers its waters at the base of Blackrock Mountain and Little Green Mountain and flows north around the Devils Elbow. Through Tanasee Lake, the river turns northwest and flows through Bonas Defeat, one of the wildest, most dangerous and least visited areas in the region. The Tanasee hydroelectric dam may release turbulent waters through the gorge without warning. Navigating slippery, bus-sized boulders takes agility, balance and strength. No marked trails exist. A fall would be disastrous.

As part of the $13 million purchase of the Balsam property, the USFS acquired the Bonas Defeat tract from James ("Jamie") McClure Clarke (1917–1999). His wife, Elspeth McClure Clarke, had inherited the property. Clarke graduated from Princeton University, served in the U.S. Navy in World War II and married the daughter of James ("Jim") McClure of Fairview, North Carolina. After growing up in Lake Forest, Illinois, and

A HIKER'S PARADISE

By Jeff Clark, creator of Meanderthals Hiking, an Internet resource for trail information in Western North Carolina, www.internetbrothers.org.

My favorite place to hike in Nantahala National Forest is Panthertown Valley, located just a few miles north of Lake Toxaway in Transylvania County. I like it for the challenges and adventure of exploration, the soft trails with their cushion of pine straw and the myriad unusual vegetation and wildlife. Large granite domes rim this picturesque valley, which is home to the headwaters of the Tuckasegee River. With more than ten thousand acres of protected land, at least two dozen waterfalls, thirty miles of maintained trails and twice that many "Carlton Trails," Panthertown Valley is a hiker's paradise.

Relating the history of Panthertown Valley is simply not complete without mentioning Carlton McNeill, "the caretaker of Panthertown." McNeill loved this forested bowl as much as he did his own family. For the last two decades of his life, he dedicated hours on a daily basis to trail building. He followed creek banks to nearly unknown waterfalls. He expanded game trails to the most stunning mountaintop vistas, leaving a legacy of love.

More than one hundred inches of annual rainfall swell the churning waterfalls along Greenland Creek and have created mountain bogs rich with plant diversity. Hiking in Panthertown Valley has increased in popularity in the twenty-first century. There is new beauty seemingly around every corner.

attending Yale University, Jim McClure (1884–1956) pastored a church in Michigan before settling outside Asheville around 1917 on his Hickory Nut Gap Farm. Throughout his life, he was a civic leader, creating the Farmer's Federation Cooperative in 1920, serving on the State Board of Conservation and Development, directing the American Forestry Association, establishing game refuges and state-owned forests and promoting public education, along with other functions. McClure also formed the North State Corporation and invested in real estate. By the late 1920s, he owned or had interest in twenty-eight different properties, including several homes, farms and lots around Asheville; a mining company in Yancey County; acreage in the Wayehutta watershed; an interest in thirty-six thousand acres in Sherwood Forest in Transylvania County; and a share in the Bonas Defeat tract.

In 1926, four Asheville businessmen paid $250,000 ($18 per acre) for the Bonas Defeat property to create a wilderness retreat. The group also purchased John Hamp Smith's seventy-acre farm above a dam and mill site on the Tuckasegee River. Smith received the payment of $6,000 in gold and silver coins at his request. The club planned to convert Smith's two-level home into a clubhouse. The proposed Tuckasegee Rod and Gun Club hoped to enroll one hundred members at $7,500 apiece to escape from "the strains of civilization that many men endured in their pursuit of wealth," wrote John Ager in *We Plow God's Fields: The Life of James G.K. McClure.* Jim McClure and others bought shares in the project shortly before the national economic crisis of 1930. Banks closed, property prices plummeted and the North State Corporation went bankrupt. The plans for a wealthy men's retreat in Bonas Defeat evaporated as the country entered the Great Depression. Doug Stuart, a friend and executive with Quaker Oats, bought one-half of McClure's share in Bonas Defeat to help McClure save Hickory Nut Gap Farm and financially recover. To generate income and pay for debts and taxes, the owners of the Tuckasegee tract entered into a long-term timber lease with Gennett Lumber Company, with an agreement to cut no trees within three hundred feet of the river or its tributaries.

"Some of the family's happiest times were spent at Tuckasegee, the one land interest saved from the crash," wrote Ager. On weekend outings, the McClures drove to the Tuckasegee River, where its waters "are forced through the gorges and over waterfalls to create spectacular sights....The churning, tumbling waters fall from 4000 to 2500 feet on this property alone...tributary streams come rushing down the steep sides of the gorge. The lower gorge, called the Shut In, because it is so difficult to pass through, begins where Bonas Defeat Cliff rises 500 feet from the stream bed."

After more than fifty-five years of the private ownership, the USFS now manages the magnificent natural wonders of Bonas Defeat Gorge as part of the Roy Taylor Forest. Nearby, a tributary, Wolf Creek, also funnels through a jumbled gorge of polished potholes, pools, fissures and waterfalls.

WAYEHUTTA OFF-HIGHWAY VEHICLE TRAIL SYSTEM

"Come enjoy fun on wheels," invited the 1995 newsletter of the Smokey Mountain Off-Road Vehicle Club (SMORVC). During the first weekend in June, the group planned to celebrate National Trails Day at the USFS Wayehutta Off-Highway Vehicle (OHV) Trail System in the Roy Taylor Forest near Cullowhee. To unite trail users across the country, three thousand other organizations were also celebrating the event on their own recreational trails.

Five years earlier, OHV riders met with the USFS district ranger to propose a new all-terrain vehicle (ATV) course in the Wayehutta Basin. Many had been riding Mead Paper Company's former logging and railroad grades before the USFS purchased the property.

"How do we start an ATV course like the Brown Mountain trail system in Pisgah National Forest and the Tellico course in Nantahala's Tusquitee District?" they asked.

"Well, you need to get organized," said the Forest Service, "and choose a spokesperson."

"That's me," said Johnny Shields, SMORVC president for twenty-six years. "The group met, discussed ideas, went back to the ranger station and asked, 'What's next?'"

"You'll need to map it out," said the Forest Service. "We don't exactly know what trails are out there yet."

"That weekend, a group of us scouted the area on ATVs," continued Shields. "We mapped about twenty miles of trails with altimeters and quad maps. We didn't have a GPS or cellphone back then." They returned to the USFS on Monday.

"You did all that in one weekend?" said the Forest Service. "You boys are serious about this."

For the past two decades, in a partnership with the USFS, the SMORVC has raised funding and donated thousands of hours of labor for Wayehutta's OHV trail system. Volunteers helped build the parking lot, comfort station and picnic shelter. They helped design the trails, clear brush, build bridges, place culverts and create water barriers. With the help of a SWECO-480 trail bulldozer and a track hoe, they continue to maintain the course today within USFS regulations. Silt fences, culvert screens and hay bales nearly eliminate erosion and silt in streams to protect the watershed.

"Over the twenty years, we've learned a lot about maintaining and designing trails," said Shields, "and we've built a trust and relationship with the Forest Service."

In the 1990s, SMORVC proposed an expansion with more miles to ride. A USFS public comment meeting erupted into a heated debate. Some organizations, like the NCWRC, voiced concern about the effects of noise and excessive human presence on wildlife habitats, especially the black bear. Sportsmen with grouse and turkey organizations felt that ATVs would disturb nesting seasons. Trout fishermen worried about muddy waters. Eventually, officials denied the request for a longer track.

Today, about 23 easy-to-difficult miles challenge Wayehutta ATV riders down dips, around curves and up steep, rocky hillsides, still maintained using environmentally sound practices by volunteers in collaboration with the USFS. Regulations require spark arrestors and sound control measures. "Get the quietest bike you can," advocated one ATV organization, "and make it quieter." On federal lands, OHV decibel regulations limit sound perception beyond a quarter of a mile. At Wayehutta, families pass a thirty-foot waterfall and ride on the easy 1.2-mile Sunset Trail for views of the Blue Ridge Parkway.

STANDING INDIAN AREA

In November 2015, a potter from Robbinsville descended the wooden steps at Winding Stair Gap on US 64 about eight miles west of Franklin. An A.T. segment-hiker named Evenflow had been hiking for about five days. Hiking northward, he had passed several southbound thru-hikers, like the seventy-eight-year-old Last Chance, planning to summit Springer Mountain, Georgia, before the winter season.

That morning, Evenflow left Deep Gap at the base of Standing Indian Mountain to hike to Wayah Bald Shelter for the night. A friend planned to pick him up at the Nantahala Outdoor Center on the following day. How many total miles of A.T. segments had he hiked? Did he keep a journal of his experiences? Did he have a goal-date to piece all of his segments together and celebrate the completion of the two-thousand-plus miles of the A.T.?

"I'm not a purist," he said with a soft grin and casual demeanor. "I don't have to touch every blaze. I'm a be-here-now hiker. I live by the Rachel Carson theory and hike for the moment. The trail calls you. Not everyone hears it, but if you do, you know it's time to go clear your brain again. The A.T. is my anti-depressant. I hike for the smiles, not the miles." With a light step and an even lighter heart, he beamed and waved as he crossed the state highway.

Driving past Winding Stair Gap on US 64-West, a motorist spots an intersection with NC 1448 (Old Murphy Road) that leads to FSR 67 and the Standing Indian Campground. The driver continues west on US

64 to explore Deep Gap, the closest point to Standing Indian Mountain accessible by vehicle. Past a bridge over the Nantahala River at Rainbow Springs, the highway crosses the boundary of Nantahala and Tusquitee Ranger Districts. Within minutes, a turn onto FSR 71 follows the eastern foot of Chunky Gal Mountain toward Deep Gap. Rainy weather and logging trucks can cause muddy ruts and potholes along the winding route. For A.T. hikers or backcountry drivers, though, this Forest Service road is worth the occasional muck.

Little Buck Creek, with mossy rocks and leafy borders, gurgles alongside the roadway. A hawk flies low in easy view. A pileated woodpecker drums steadily on a dead snag, waiting for its mate. A fallen leaf held by an invisible strand of spider's silk twirls in midair like magic. The Park Ridge Trail connects the Forest Service road to the Standing Indian Campground. At Deep Gap, shuttles pick up A.T. thru-hikers needing supplies in Franklin. Other hikers need return transportation to a trailhead. Some hike the Kimsey Creek Trail past gardens of spring wildflowers to the campground. In June, others climb the steep trail up Standing Indian Mountain past gorgeous purple rhododendron. Return to US 64 East and take the road to the Standing Indian Campground.

In 1928, when the campground was a logging camp, six-year-old Marie Cloer and her five siblings moved into one of the portable houses at White Oak Bottoms. W.M. Ritter Logging Company had hired her father, George Cloer (1884–1970), as camp preacher and blacksmith. Wide, rough boards covered the exterior walls of the company housing, and slender boards sealed the gaps and tarpaper protected the rooftops. When Ritter relocated to new logging sites, crews attached a cable, inserted through a hole in the home's roof, to a floor sill. Steam-powered log loaders hoisted the homes onto flatbed railway cars and transported them to the next camp. The furniture, moved to the center of the room, helped balance the twelve-foot-wide structures.

For about five months a year, Marie attended a camp school with other first- to seventh-grade students in a one-room schoolhouse. When Ritter moved its logging operations two miles south of White Oak Bottoms to Frog Flats at the confluence of Little Indian Creek and Nantahala River, the little schoolhouse went, too, using the same cable and log loader system. Later, its schoolteacher, Mattie Brendle Fouts, helped shape the young minds at the final logging camp about six miles farther south. Near the juncture of Big Indian Creek and Nantahala River, Fouts received a salary of seventy-two dollars to serve as its last teacher between 1935 and 1936.

W.M. Ritter Lumber Company's Shay locomotive moves the company's temporary housing units to a logging camp. *Macon County Historical Society and Museum.*

At White Oak Bottoms, Marie's father held Sunday services at the schoolhouse. On Wednesday nights, in the communal lobby railroad car, he preached to loggers lining the long benches on each wall. In an interview with USFS Bill Lea, Marie said the loggers "chewed tobacco and spit in spittoons while Daddy preached." Other camp railroad cars included a combined store and post office, a kitchen, a dining car and sleeping quarters for unmarried men. Sixteen to twenty loggers earning one dollar a day slept on bunkbeds in unheated railcars. Sixty-gallon barrels held water for bathing. A separate car accommodated the camp foreman, timber-cutting foreman and office staff.

Nearby, a teamster's shack housed the horseman who tended the massive draft horses. Each day, he harnessed the teams, often attaching an oxen tail for decoration. Teams of dapple-gray, 1,800-pound Percherons, hitched two abreast, could skid about 2,500 board feet of logs for more than a mile. With hammer, heat and anvil, Preacher Cloer made horseshoes and logging equipment. When Marie visited her father's forge after school, he created tiny knives from ten-penny nails to cut mud pies in her playhouse. She asked for sharper ones, but he refused.

Daily trains arrived from Ritter's band sawmill at Rainbow Springs four miles away, passing Cloer's blacksmith shop—located, presumably, at the present campground host site. Cloer visited the sick and attended funerals, driving one of the few cars available along a rough road from the logging camp to Wallace Gap, now part of FSR 67. An emergency phone call from the camp store could summon Dr. McGuire at Rainbow Springs, making his rounds on a four-person railcar.

In 1927, when Ritter Lumber Company was operating about eleven sawmills in six states, it had moved its logging equipment from Proctor in the Great Smoky Mountains to Rainbow Springs. C.W. Slagle, a local surveyor, owned the wide field and leased it to Ritter to build his band sawmill and mill town. In September 1928, Ritter's newsletter, the *Hardwood Bark*, stated, "High in the hills, we rode through an indescribably beautiful range of mountains to a little town as pretty as its name, Rainbow Springs…six or seven blue imposing mountains form a basin in which the village nestles comfortably. Through the bottom of this natural bowl courses the cold swift stream of the Nantahala."

Trains transported logs from White Oak Bottoms to Ritter's three-story band sawmill. Damming the Nantahala River, the company created a 180- by 100-foot pond beside the mill. Loaded flatbed railcars rolled in beside the lake. Men unbound huge chains, releasing the heavy load into the seven-foot-deep waters. Washed of dirt and debris, the logs traveled by conveyor belts into the mill. Fueled by burning wood scraps and coal, the steam-powered mill could process 50,000 board feet per day, enough wood

W.M. Ritter Logging Company operated a large band sawmill at Rainbow Springs. *Macon County Historical Society and Museum.*

The USFS converted a Model T truck to deliver supplies and watch for wildfires along some logging railroads. *National Forests of North Carolina Historic Photographs, D.H. Ramsey Library, Special Collections, University of North Carolina–Asheville.*

to furnish the framework material needed to build three 2,400-square-foot homes. Lumber, stacked on hickory timbers, air-dried out front. Ritter's railroad transported the processed lumber north along Nantahala River past Aquone to a depot near Hewitt in Nantahala River Gorge. Hooking up with Southern Railway's Murphy Branch, the lumber moved to western markets beyond Andrews or eastern markets beyond Bryson City. After placing railroad wheels on a Model T Ford truck, the USFS traveled Ritter's rail lines to transport supplies and monitor for wildfires.

At Rainbow Springs, unmarried men lived in a twenty-six-bedroom clubhouse with dining room, kitchen and showers. A boardinghouse sat on the hill, and company housing dotted the fields. The mill town had a commissary, a tool house, a post office, a blacksmith, a barbershop and a medical facility. At the corner of present-day US 64 and Old Murphy Road, a five-teacher schoolhouse served the mill town children. Alongside Old Murphy Road, a former camp that housed convicts building Macon County roads became extra shabby-shack homes for loggers.

In 2001, Lea interviewed Siler Slagle, C.W.'s grandson. After Ritter closed Rainbow Springs during the Great Depression, said Slagle, the company borrowed money from local employees to reopen the business. Locals had not deposited their personal cash in banks that later failed. By 1939, Ritter had moved its operations from Rainbow Springs to Hayesville, North Carolina, leaving behind stacks of finished lumber. A small engine moved it to the road, where buyers trucked it to other sites. Dismantled mill houses sold for fifty dollars. Since C.W. Slagle had died in 1931, the mill site cleanup fell to his sons, Carl and Burt. In 1941, Burt removed the rest of the mill and recycled some of its equipment in a local truck garage. Carl dug up the hickory timbers that held the stacks of lumber and filled in the lake. Today, a brushy green pasture alongside US 64 erases its former days as a busy mill town.

The White Oak Bottoms logging camp became the USFS Standing Indian Campground. Legend says that an Indian warrior, stationed on Standing Indian Mountain to guard children from a winged monster trying to steal them, fled when the beast appeared. The Great Spirit turned the Indian to stone and killed the monster with a bolt of lightning.

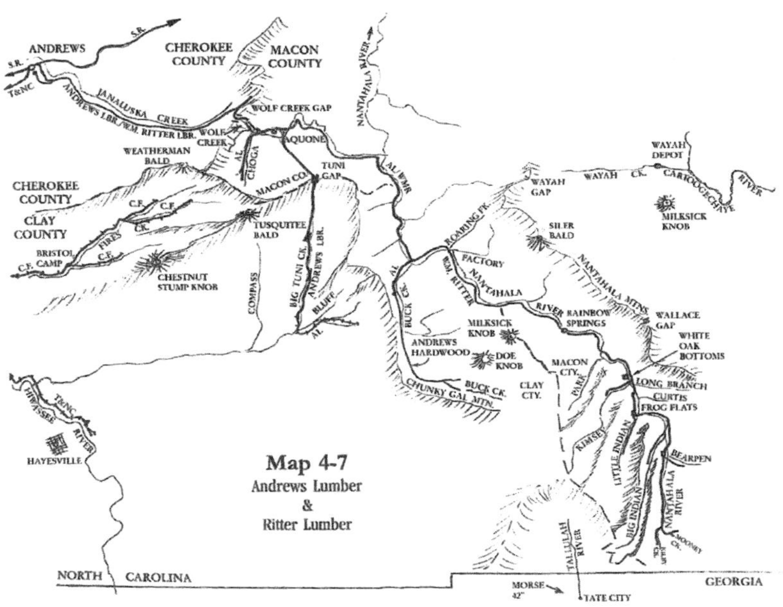

Railroads in the Nantahala region. *Permission to use from Thomas Fetters,* Logging Railroads of the Blue Ridge and Smoky Mountains, *vol. 2.*

Fire lookout cabin on Standing Indian Mountain. *National Forests of North Carolina Historic Photographs, D.H. Ramsey Library, Special Collections, University of North Carolina–Asheville.*

The Cherokee name for the mountain means "where the man stood." On Standing Indian Mountain in the 1920s, the USFS stood on a cabin rooftop watching for wildfires. By 1929, the agency built a wooden lookout tower on a stone foundation but removed it in the 1950s. Straddling the Continental Divide, the 5,499-foot summit guards the headwaters of Tallulah River flowing south to the Savannah River. To the north, it gives rise to rivulets feeding the headwaters of the Nantahala River.

Five-thousand-foot peaks nearly surround Standing Indian Campground. Campers drive past the camp store and the first loop of campsites and cross the Nantahala River to enter the main campground. Kimsey Creek, a tributary of the Nantahala, flows through the camping area. Four more camping loops offer open green meadows with views of the mountains or secluded, forested sites for privacy. In August 1983, Bill Sargent, an *Asheville Citizen-Times* journalist, received a tip. While he and his family were visiting the Wilson Lick Ranger Station, USFS volunteer Richard Byrd discussed the historic cabin. When asked about other significant NNF sites, Byrd replied, "The Standing Indian Campground is regarded as the finest USFS campground in the country." Sargent drove twenty minutes west of Franklin to "one of the few campgrounds that I've ever seen where you have trouble getting your feet dirty."

In the mid-1980s, the USFS renovated the campsites. Employees with the Senior Community Service Employment Program (Older Americans Program) did most of the work. Many of the local workers were skilled rock masons and woodworkers. One older American had lived there when it was a logging camp. In 1926, when five-year-old Bessie Bingham Stockton's mother died, she moved to White Oak Bottoms to live with her lumberjack brother and his wife. While helping restore the campground, the sixty-three-year-old took a nostalgic stroll to her old cabin site, now the camp's amphitheater. The spring that had supplied their water was still flowing nearby.

Departing Standing Indian Campground, visitors return to FSR 67. At a small parking space to the left, a 1.4-mile round-trip trail through a high-elevation cove hardwood forest leads to the John Wasilik Memorial Poplar tree. The A.T. passes near the trailhead. In 1969, the USFS placed a plaque in a boulder near the tree, dedicating the country's largest poplar to the NNF ranger. Later that year, officials learned that the eight-foot-diameter column

Employees with the Older Americans Program assisted the USFS with many projects.
Lewis Kearney.

Ranger John Wasilik is stamping "U.S." on logs in NNF. Logs cut on government land are measured, then the operator is charged its net volume. *National Forests of North Carolina Historic Photographs, D.H. Ramsey Library, Special Collections, University of North Carolina–Asheville.*

was second in size to one in Virginia. In the 1930s, loggers cut a tree of equal size, thought to be 250 to 300 years old, that oxen struggled to remove, so the men left this one.

USFS supervisor Peter J. Hanlon dedicated this monarch to "an outstanding forester. He knew that by protecting the vital resources—water and soil—all other benefits of a forest can be retained and even multiplied." USFS deputy regional forester H.C. Eriksson said, "This outstanding tree specimen represents the strength and ruggedness for the outdoors which characterized John so well."

About ten years later, lightning struck the giant tree, and it began to die. By 2011, the USFS had begun warning visitors to stay at least fifty feet away from the trunk. "The great Wasilik poplar's time at the top may be numbered," wrote Quintin Ellison in the *Smoky Mountain News.* Today, its fallen, decomposing trunk replenishes the ecosystem with valuable nutrients and food sources for the next generation.

A right turn out of the Standing Indian Campground onto FSR 67 leads to the Hurricane Horse Camp, where equestrians explore twenty miles of woodland trails. A great fall color drive, the graveled road tightly winds past trailheads to Big Laurel Falls, Mooney Falls, Albert Mountain Fire Tower and Pickens Nose before it exits the forest at the USFS Coweeta Hydrologic Laboratory.

South of Standing Indian Mountain, Congress created the odd-shaped Southern Nantahala Wilderness Area in 1984. Crossing the state line, it lies in Chattahoochee National Forest in Georgia and in the Tusquitee and Nantahala Ranger Districts of NNF in North Carolina. The USFS bought the North Carolina wilderness acres from Ritter Lumber Company, but Ritter retained the timber rights for twenty years, agreeing to cut no trees less than fourteen inches in diameter. Deep in the wilderness, on the banks of the Tallulah River, the Dorothy Thomas Foundation had owned a two-thousand-acre Girl Scouts Camp since 1963. In 1984, it agreed to sell the acreage to the USFS if the agency would protect it as part of the designated wilderness area. The government's purchase saved the Girl Scouts' five-thousand-foot peak, Big Scaly, from a developer wanting to build a ski lodge.

Christi and Tim Worsham scan the horizon from Pickens Nose. *Photo by author.*

In Chattahoochee National Forest, the USFS acquired the wilderness tracts from private landowners, investors, developers and the Nature Conservancy.

In the 1980s, Taylor Crockett, a former logger with Ritter Lumber Company, became a conservationist, campaigning for federal wilderness protection of the southern Nantahala region. "We asked for 40,000 acres," he said, so the outcome was disappointing. However, the nearly twenty-four-thousand acres today protect numerous waterfalls and creeks; special plants, such as the large purple-fringed orchid and granite dome bluets; and four-thousand- to five-thousand-foot summits. Old-growth forests of yellow birch and northern red oak thrive on steep mountainsides.

COWEETA HYDROLOGIC LABORATORY

East of the Southern Nantahala Wilderness Area, the USFS Coweeta Hydrologic Laboratory conducts research on the effects of forest use and management practices on watershed ecosystems. In 1934, Charles R. Hursh, researcher for the USFS Appalachian Forest Experiment Station in Asheville (now Southern Research Station), chose the Coweeta drainage basin in Nantahala National Forest for a permanent research lab in watershed science. Within the 5,400-acre Coweeta watershed, many streams form smaller watersheds and flow into Coweeta Creek, forming the headwaters of the Little Tennessee River. With a well-defined basin containing many streams and land gradients, a variety of soils, a rich biodiversity and an annual rainfall from seventy to ninety inches, Coweeta provided the perfect location for the study of soil and water. In 1948, the Coweeta Experimental Forest became the Coweeta Hydrologic Laboratory.

Operating a sawmill at Otto, North Carolina, Ritter Lumber Company owned and logged the Coweeta area before selling it to the USFS. When the Forest Service chose Coweeta as the basin for its new watershed research center, Civilian Conservation Corpsmen from Camp F-23 constructed a three-mile access road in 1935 from Otto into Coweeta. In addition to working in Nantahala National Forest, the CCC helped survey Coweeta's boundaries, develop a nursery, inventory permanent vegetative plots, build three trails with sixty rain gauges and construct a chestnut log office building, bunkhouse and residence. World War I–era and other government surplus supplies equipped offices and machine

Left to right: Rex Duvall, Lloyd Parrish and Ray Welch, all of Franklin, served with the CCC at Camp Otto. *Ray Welch and family.*

shops and provided pipes, valves, pumps and other items for the water measuring devices.

"I had to clear all the trees off a hillside at Coweeta," said ninety-one-year old Ray Welch, a former CCC enrollee at Camp Otto. "You couldn't leave any brush behind; it had to be as clean as this floor." Sitting in his home ten miles north of Franklin, built with CCC earnings sent home to his father, Ed Welch, Ray spoke with pride about his work with the Corps.

"We cleared that hillside at Coweeta in January, when it was bitter cold," he said. "We worked unless it got more than ten degrees below zero. Other men in my troop built a dam for the Coweeta lab equipment to determine streamflow."

Hard work and tough times were early childhood lessons. Born in 1924 near Burningtown Creek, Ray walked to school to fire up the wood heater in the two-room schoolhouse and "chop ice out of the sink." "It's no wonder the school was so cold," he explained. "The cracks were so wide that a cat could walk through." Earning fifty cents a day for eight hours' work, twelve-year-old Ray and a mule plowed steep hillsides for the neighbors. At fourteen, he hitched a ride with other young men from Franklin to work at a sawmill in Georgia.

"I was as strong as an ox," said Ray. "I could carry a hundred-pound sack up a hill. So I fibbed about my age." At the sawmill, he earned twenty-five cents an hour and paid seventy-five cents a day for room and board. When he heard that he could make thirty dollars a month, sending twenty-five dollars of that home to help his family, he lied about his age again and signed up for the CCC. In 1939, he and about two hundred other enrollees went to work at Coweeta.

CCC camps rewarded hard work with recreational activities, such as baseball, boxing and music. In *That Magnificent Army of Youth and Peace*, Harley Jolley mentioned that thirty CCC recruits at Otto performed in the play *Murder in the CCC Camp*. Cast members from the Federal Theatre Players of Atlanta joined them in the performance on August 11, 1937.

"I had a car," said Ray. "It was against the rules, so I hid it in a farmer's field." When the movie *Gone with the Wind* was first released, he and some of the boys piled in his car to drive to town to see it. When the CCC offered him a job out west, he "jumped on it." He was the only one at Otto at the time to take advantage of that opportunity. He boarded a train at Otto for Fort McPherson, Georgia, and onto Kelso, Washington. After discharge from the CCC as a night watchman and truck driver, he became a welder at a shipyard and served in the navy during World War II. In 1946, when there were no jobs in Franklin, he moved to Michigan and worked for Pontiac Motors. Retirement found Ray back at his boyhood home in Welch Cove near Burningtown Creek.

At Coweeta, the CCC helped build nine weirs for scientists to collect data on the effects of logging, farming and woodland grazing on valuable watersheds. Perfectly engineered to contain the water flowing through a watershed, concrete channels, or weirs, funnel water across a knife-edge weir blade held in its concrete headwall. Anchored to the stream's bedrock, the headwall spans the width of the watershed. An adjoining gauging station automatically records the volume of water passing over the blade. The water flow height is later converted mathematically into streamflow.

Since 1934, streamflow gauges at sixteen different sites have continuously measured the stream water discharge off a watershed every five minutes. Weather stations and rain gauges have recorded temperature, humidity and precipitation. During World War II, when CCC servicemen joined the military, "the loyalty and hard work of local workers kept Coweeta alive during the war years," wrote Wayne Swank, retired Coweeta scientist and

The CCC helped build streamflow measuring devices at the USFS Coweeta Hydrologic Lab, like this V-notch weir, flume and gauge house. *National Forests of North Carolina Historic Photographs, D.H. Ramsey Library, Special Collections, University of North Carolina–Asheville.*

coauthor of *Forest Hydrology and Ecology at Coweeta.* In less than six hours, three local men read all the rain gauges along thirty miles of trails after every storm (about every third day), "a job that once took twelve CCC corpsmen!"

After scientists gathered baseline streamflow data on forested watersheds in their natural state for six years, they began field experiments on individual plots to determine how logging, grazing, forest cover changes and other disturbances affected the quality and quantity of water in a watershed. Researchers clear-cut some of the experimental watershed plots. They planted some with corn and others grasses, demonstrating that dense grass used about the same water as a deciduous forest. Over time, loss of topsoil on the hilly terrain limited corn production.

In one experimental plot, cattle grazing on mountainsides beneath hardwood trees eventually compacted the soil, increasing the water runoff. On two watersheds, scientists converted hardwood forests to white pine plantations with surprising results. Since the evergreen pine transpires all year (loses water to the atmosphere through the surface of its leaves or needles), an acre of white pines used 250,000 gallons more water per year than the hardwood trees, which stop transpiring when they drop their leaves in the fall. Thus, white pines reduced the streamflow in a watershed.

Other experiments have addressed the watershed effects of timber-harvesting methods, such as clearcutting in strips, thinning an understory and cutting mixed-hardwoods on north-facing slopes verses south-facing ones. After studying the amount of vegetation removed and the amount of solar energy reaching each slope, scientists learned that the streamflow after the harvest of north slopes was twice the flow of the logged southern slopes. Researchers at Coweeta have also studied the watershed effects of the design of Forest Service roads, the harvesting of timber from riparian areas and the practice of prescribed burning. They have conducted studies on groundwater movement and litter decomposition. Soil scientists collect weekly soil samples in the basin for nutrient analysis. Shallow lysimeters sample the soil nutrients available to plants; deeper ones penetrate beneath root systems where the nutrients will potentially join the water systems. At Coweeta, scientists have found that managed forests with deep, rich soils absorb up to four inches of rainfall per hour. Denuded lands with eroded topsoil absorb less than one-fifth of an inch of rain per hour.

In 1960, Congress passed the Multiple Use–Sustained Yield Act to protect the timber, grazing, recreational, wildlife and water resources of the nation's forests. In response, Coweeta conducted a crucial long-term study to address multiple-use and sustainability issues.

In 1952, a 3-D model of the Coweeta Hydrologic Laboratory, created for the film *Waters of Coweeta*, was later displayed at a National Watershed Meeting in Washington, D.C. Pictured here in the Department of Agriculture lobby are, *left to right*, Richard E. McArdle (*far left*), chief of the forest service; a staff member under the secretary of agriculture; Ezra Taft Benson (*pointing*), secretary of agriculture; and President Dwight D. Eisenhower. *USFS.*

"It provided the guiding philosophy for the management of national forests," a former Coweeta scientist, Lloyd Swift, stated in the film *The Story of the Coweeta Hydrologic Lab.* "When the Environmental Protection Agency said you need to have a code of best management practices for the forests you manage, the states came to Coweeta to see how to do it."

Today, land managers, engineers, conservationists, city planners and students from around the world come to Coweeta in NNF to learn how to manage forest ecosystems to ensure sustainable water resources. In 1980, Coweeta Hydrologic Lab began its participation in the Long Term Ecological Research program. Funded by the National Science Foundation and hosted by the University of Georgia, Coweeta is now one of twenty-six sites in the network. A current focus is to understand the region's past two hundred years of land-use changes to help plan for expected patterns in the upcoming thirty years. Coweeta scientists are also focusing on current concerns, such as climate change, acid rain and invasive species.

In 1995, research studies began at the 1,300-acre USFS Blue Valley Experimental Forest, south of Highlands near the Georgia border. Recently, Coweeta scientists studied the effects of increased sunlight on hemlock trees infected with the hemlock woolly adelgid. At Coweeta and Blue Valley, the staff established hemlock study areas and cut 0.3-acre sunlit gaps around the trees in half of each plot. Researchers treated the Blue Valley hemlock stand, but not Coweeta, with predator beetles. The results showed that by using the combination of sunlight and predator beetles, forestry managers may improve the long-term survival rate of hemlocks.

Understanding the effects of natural and human disturbances on a watershed and its ecosystem, and the relationships between atmosphere, environment and forest watersheds, helps land managers make wise decisions. Basic knowledge starts with understanding the hydrologic cycle in a forested landscape. Water falls from the atmosphere as precipitation and soaks a forest canopy. After canopy leaves become saturated, water droplets fall to shrubs, flowers and other plants in the understory. Water runs down plant stems, drips from leaves or falls directly to the earth's soil to replenish plant and animal life. At Coweeta, like other areas that annually receive seventy inches of rain, thirteen of those rainfall inches never seep into the soil. They are intercepted, instead, by the leaves of plants and the leaf litter on the forest floor. After photosynthesis, plants return twenty-two inches to the atmosphere through evapotranspiration. The remaining thirty-five inches, which equals about half the annual rainfall at Coweeta, seeps into the subsoil to enter streams and eventually evaporates back to the atmosphere. High schoolers learn the hydrologic cycle, but at Coweeta Hydrologic Laboratory, one of the oldest continuous environmental labs in the world, the principles of the hydrologic cycle are studied in relation to forest management practices to ensure a healthy future for watershed resources.

NANTAHALA RIVER GORGE

Nantahala River used to flow right here," pointed Payson Kennedy, co-founder of Nantahala Outdoor Center (NOC).

Sitting on the twenty-foot-high deck outside his home, Payson explained that the Tennessee Valley Authority (TVA) had changed the course of the Nantahala River in the 1940s. Since the late 1800s, Southern Railway's Murphy Branch rail line had crossed a wooden bridge at each end of a U-shaped bend in this section of the river. During the war years, when steel and lumber were scarce, TVA blasted the river at the oxbow to straighten its course and moved the bridges for its own railroad at TVA's Fontana Dam construction site. Rocky debris from the blast dammed Nantahala's natural riverbed. That jagged debris, as well as the sharp remnants from the construction of the new route of the Murphy rail line, created a twelve-foot drop in Nantahala's new course. Now known as Wesser Falls, these rapids are located about one hundred yards downstream of NOC's kayak training gates.

In 1971, Payson and Aurelia Kennedy and their four children moved from Atlanta to the banks of the Nantahala River. A friend and fellow canoeist, Horace Holden, had purchased the Tote 'n Tarry Motel and gas station in the gorge. They wanted to establish a whitewater recreational center. To manage the business, Aurelia left a teaching position and Payson left his job as a librarian for the Georgia Institute of Technology.

"I wanted a career change," he said. "I wanted to get outdoors and to be with people who loved to paddle. I had talked to the North Carolina

Outward Bound about starting a whitewater training program but decided to help Horace start the business here instead."

In 1942, Payson learned how to canoe as a Cub Scout. During the summers of 1950 and 1951, he worked as a camp counselor, which included canoeing instruction at the Atlanta YMCA's Camp Pioneer. Nineteen-year-old Aurelia (Relia) had taught canoeing at Camp Merrie Woode in Sapphire, North Carolina. In 1954, Ramone (Ray) Eaton gave Relia a wedding present: Ray and Relia paddled a tandem wood and canvas canoe over the Class III Nantahala Falls on the Nantahala River. Ray Eaton and "Chief" Frank Bell were Western North Carolina whitewater pioneers.

"I'm often introduced as Frank Bell's grandson," said Will Leverette, whose mother, Pat Bell, raised him paddling on the Chattooga, Green and Nantahala Rivers. In 1922, his grandfather started Camp Mondamin, a boy's camp, near Tuxedo, North Carolina. The next year, Frank Bell began canoeing instruction, taking twelve campers to the Mississippi River. Will's mother was the first director of Camp Mondamin's sister camp, Green Cove

Ray Eaton navigates Nantahala Falls in the early 1950s in his aluminum Grumman Canoe. *Will Leverette.*

Camp, considered today as one of the world's finest whitewater training camps for girls. After studying art at New York University in the 1950s, Pat Bell married a professor of history and moved to South Carolina, camping and paddling on Green River on family vacations.

"In 1945, my grandfather's wife, Cala, went to visit family in Bryson City," said Will. "She learned that a power company had built a new dam across the Nantahala River." When she returned home to Camp Mondamin, she told camp leaders about a potentially new river to paddle. Ray Eaton, camp paddling instructor and later senior vice-president of the American Red Cross, scouted the Nantahala area alone and soon returned with five camp leaders. Ray, Pat Bell, the "Chief" and three others—Fritz Orr Sr. (who later established Camp Merrie Woode), John DeLabar (paddling master for whom DeLabar's Rock in Nantahala River is named) and Billy Pratt—ran the river in wooden canoes, thought to be the first paddlers to run Nantahala's rapids. Pat wanted to paddle down Nantahala Falls on that trip but followed her father's wishes and portaged around them.

"Those camps, providing paddling instruction on the Nantahala and other area rivers," said Will, author of *A History of Whitewater Paddling in Western North Carolina*, "have been the frontrunners for training world-class champions. Camp Mondamin has trained about twelve Olympians." Will is a volunteer paddling coach for Warren Wilson College and for the local chapter of Team River Runner USA, a whitewater program for disabled veterans—"the most rewarding, stimulating, invigorating, worthwhile paddling instruction I've ever done."

By 1972, Holden and the Kennedys had opened NOC and joined business with pleasure on the Nantahala River. That year, *Deliverance*, a movie filmed on the Chattooga River, introduced a wider public to whitewater activities. During the filming, Payson was the paddler-stuntman double for Ned Beatty, but "very little of the footage of the doubles was used in the final version of the film," said Payson. Also in '72, when the Olympics Committee recognized canoeing and kayaking as Olympic sports, the new NOC began offering canoeing instruction. At that time, it was the only commercial whitewater outfitter in the area. It had the capacity to run only one raft trip a day on the Chattooga or Nantahala River. By the early 1980s, thirteen companies offering rafting trips on the Nantahala had opened.

"We needed some sort of regulation," said Payson. "We saw increased streambank erosion and sanitation problems. People were camping everywhere along the river. When too many people used the river at one time, you got in each other's way."

TRAINING FOR THE OLYMPICS ON THE NANTAHALA RIVER

By Joe Jacobi, Olympic gold medalist, author, professional speaker, performance strategist and executive trainer

To perform your best at the Olympic Games is a resource-driven pursuit. Many elite athletes talk about the need for money, gear and coaching to support their quest for gold, but before you figure out what you need, you must understand what you already have.

As our home and primary training site leading up to the 1992 Olympic Whitewater Canoe Slalom events in Spain, the Nantahala River was our fortress of simplification. The river moved my training partners and me away from the non-essential, allowing us to discover exactly what was important. Training at the Nantahala afforded us the gift of time. The forces of nature rallied us to stay out on the water a little bit longer and confront the boundaries of fatigue and focus. In the process of challenging ourselves and each other on the water, the Nantahala became the bond of lifelong friendships off the water. The confluence of nature and community instilled a desire to learn, innovate and, ultimately, serve something bigger than ourselves. Our small section of the river, just beside the Nantahala Outdoor Center, became a gravitational point for people who wanted to lose themselves… as much as find themselves.

From its fluctuating currents to frosty water temperatures, the changing nature of the Nantahala River gave to me the foundations of a good life. Without the Nantahala River, my canoe partner, Scott Strausbaugh, and I wouldn't have won America's first Olympic Gold Medal in Whitewater Canoe Slalom; I wouldn't have met my wife; and we wouldn't have had a daughter who cut her whitewater teeth on this magnificent river. Today, finding solutions to life's problems is as easy as finding our way to the Nantahala River. Whether it's for the first time or the 100th time, the river meets us right where we need to be met—in pursuit of our life's "best moments."

Above: In December 2011, Michael Gora photographed this pair of whooping cranes in a field near the Hiwassee River in Clay County. In 1999, scientists began conditioning the endangered cranes with an ultralight aircraft to fly from breeding grounds in Wisconsin to a refuge in Florida, passing over east Tennessee. The banded male in this photo had wintered at the Tennessee's Hiwassee Wildlife Refuge for the past two years. The pair remained near Hayesville for two to three weeks. *Michael Gora, www.mikegora.com.*

Left: Kitty Myers marvels at the wonder of the Great Wall of Panthertown Valley. *Thomas Mabry, https://thomasmabry.smugmug.com.*

Wayah Bald Fire Tower. *Eric Haggart, Timeless Moments Imaging, https://www.timelessmomentsimaging.com; www.facebook.com/tmi.imaging.*

Cullasaja Falls in winter. *Eric Haggart, Timeless Moments Imaging, https://www.timelessmomentsimaging.com; www.facebook.com/tmi.imaging.*

Above: When the controlled dam at Santeetlah Lake releases its mighty roar down the Cheoah River, experienced whitewater paddlers from all over the country come to play. *Photograph by author, taken with Canon EOS-60-D, along NC 129, April 16, 2016.*

Right: Explorer and photographer Thomas Mabry took an extremely difficult hike outside Cashiers to stand in the plunge pool below Little Canyon Falls in the upper Whitewater River slot canyon. *John Forbes.*

Left: Yellow lady's slipper. *Diane Bauknight.*

Right: The intricate beauty of some plants is best seen through a magnifying lens. NCWRC biologist Gabrielle Graeter studies a bog habit in NNF, one of the rarest natural communities in North America. *Gary Peeples, U.S. Fish and Wildlife Services.*

Mountain camellia. *Angela Martin.*

Right: During a guided hike in the Balsams, hosted by the Western Carolina University Plant Conference, John Spencer and participants learned fern identification from noted author and naturalist George Ellison. *Photo by author*.

Below: After a thirty-seven-year career with the USFS Cheoah District, Hoot Gibbs volunteers to help maintain hiking trails with Buster Stewart, retired USFS Jim Buckel, Larry Icenhower and Alf, the donkey. *Hoot Gibbs.*

Left: Cain Ammons, who began fishing at age three, holds a largemouth bass recently caught at Santeetlah Lake. *Austin Ball/Lisha Ammons.*

Below: Winter sunset from Chatuge Lake from the Jackrabbit Campground. *Michael Gora, www.mikegora.com.*

Dick Evans and his grandson, Jackson Smith, hike the Yellow Creek Mountain Trail near Tapoco. *Dick Evans.*

Frozen Dry Falls. *Riley Henry.*

Right: A bugler plays "Taps" at the Memorial Service for the military men who lost their lives in a plane crash near the current Cherohala Skyway. *Kim Hainge, godsgardens@earthlink.net.*

Below: Each October and February, the afternoon sunrays cast a shadow of Whiteside Mountain, known as the "Shadow of the Bear." *Rodney Byard.*

Right: A painter captures the beauty of Dry Falls. *Photo by author.*

Below: There is a "Golden Hour at Sunset" seen over the Cowee Mountains from the Blue Ridge Parkway. *Kristina Plaas, Plaasabilities Photography.*

Right: Chaperoned by a NNF fire management officer, Smokey Bear waves to crowds lining Main Street, Highlands, in December 2015. *Photo by author.*

Below: Jackrabbit Mountain Recreation Area. *Joanna Padgett-Atkisson.*

Nature has its way of offering hope and inspiration, like this Bartram Trail view near Jones Gap outside Highlands. "It's as if the clouds were talking," said Stefanie Reponen. *Thomas Mabry, https://thomasmabry.smugmug.com.*

Moses prepares to release a banded Louisiana waterthrush. *Tyson Smith, www.wildandwonderfulphotography.com.*

Centerfold map. *Illustration, Ken Czarnomski, architect/illustrator/cartographer/naturalist, phoenix2reach@gmail.com.*

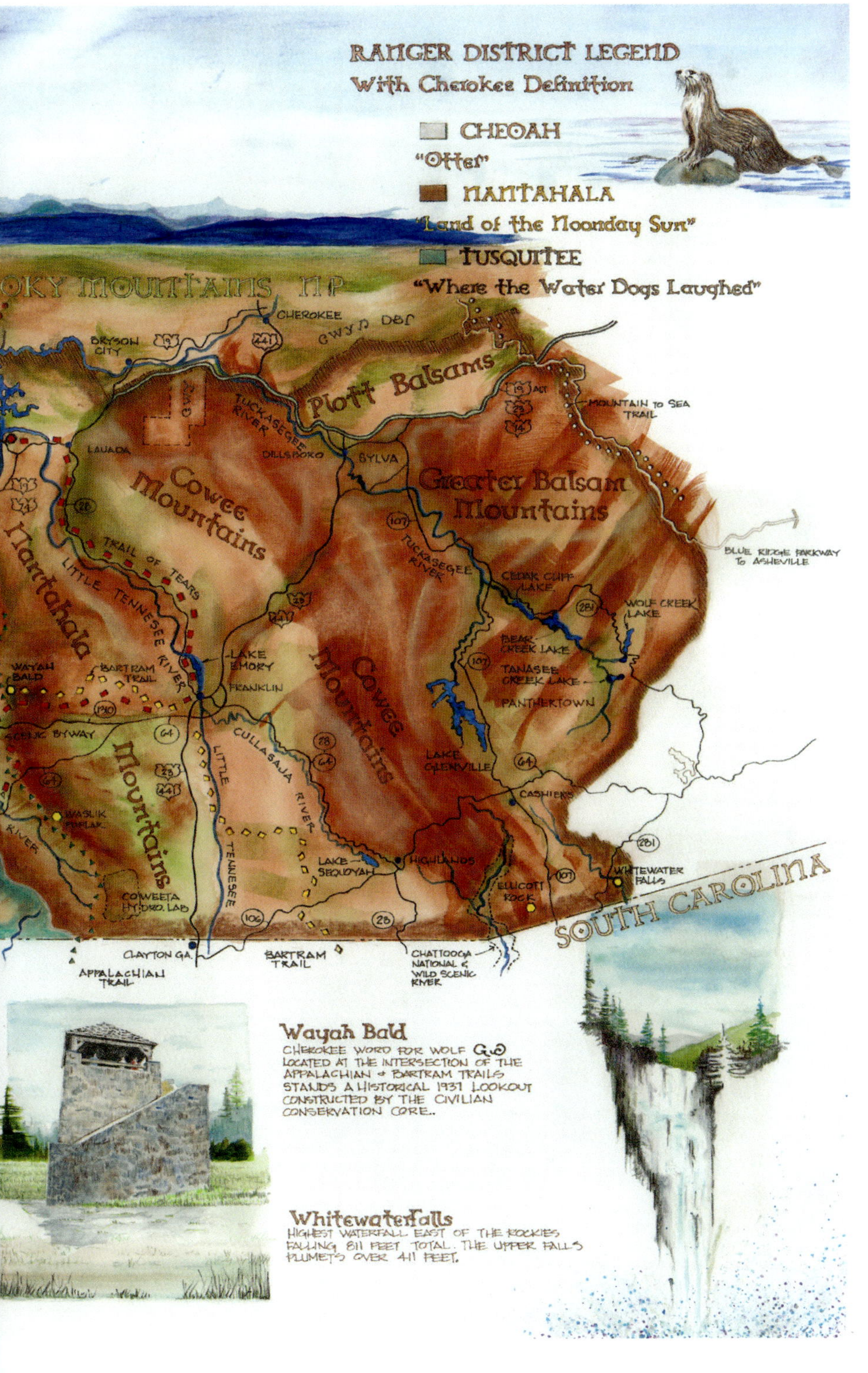
RANGER DISTRICT LEGEND
With Cherokee Definition
CHEOAH
"Otter"
NANTAHALA
"Land of the Noonday Sun"
TUSQUITEE
"Where the Water Dogs Laughed"
OKY MOUNTAINS NP
CHEROKEE
BRYSON CITY
Plott Balsams
TUCKASEGEE RIVER
DILLSBORO
SYLVA
MOUNTAIN TO SEA TRAIL
LAUADA
Cowee Mountains
Greater Balsam Mountains
BLUE RIDGE PARKWAY TO ASHEVILLE
Nantahala
TRAIL OF TEARS
LITTLE TENNESSEE RIVER
TUCKASEGEE RIVER
CEDAR CLIFF LAKE
WOLF CREEK LAKE
BEAR CREEK LAKE
TANASEE CREEK LAKE
PANTHERTOWN
WAYAH BALD
BARTRAM TRAIL
LAKE EMORY
FRANKLIN
Cowee Mountains
SCENIC BYWAY
Mountains
LITTLE TENNESSEE
CULLASAJA RIVER
LAKE GLENVILLE
CASHIERS
RIVER
COWEETA HYDRO. LAB
LAKE SEQUOYAH
HIGHLANDS
ELLICOTT ROCK
WHITEWATER FALLS
SOUTH CAROLINA
CLAYTON GA.
BARTRAM TRAIL
APPALACHIAN TRAIL
CHATTOOGA NATIONAL & WILD SCENIC RIVER
Wayah Bald
CHEROKEE WORD FOR WOLF
LOCATED AT THE INTERSECTION OF THE APPALACHIAN & BARTRAM TRAILS STANDS A HISTORICAL 1937 LOOKOUT CONSTRUCTED BY THE CIVILIAN CONSERVATION CORE.
Whitewaterfalls
HIGHEST WATERFALL EAST OF THE ROCKIES FALLING 811 FEET TOTAL. THE UPPER FALLS PLUMETS OVER 411 FEET.

Above: Yellow Creek Falls, Cheoah District. *George Evans, gwevansiii@yahoo.com.*

Left: Whitewater Falls. *Bill Lea.*

Left: Wake robin trillium. *Gretchen Kirkland.*

Middle: Thousands of firefighters helped battle the devastating 2016 North Carolina wildfire season. *USFS.*

Bottom: Whiteside Mountain. *William Nichols.*

Left: The daring adventurer, Thomas Mabry, provides a glimpse of the dangerous gorge on the Tuckasegee River, known as the Bonas Defeat. *John Forbes.*

Below: Sunset view from Yellow Mountain, Nantahala Ranger District. *Thomas Mabry, https://thomasmabry.smugmug.com.*

In the early 1980s, USFS staff collected data on the recreational use of Nantahala River. "With the increased numbers of people on the river, the agency wanted to initiate guidelines for a special-use permit system to protect the resource," said retired district ranger Lewis Kearney.

TVA furnished cameras to record people floating the river. For hours, a Western Carolina University volunteer viewed the films and recorded the gender and ages of the users. She recorded the time of day, day of the week and watercraft used. With this and other data, the USFS and the outfitters initiated a special-use permit system for Nantahala River and determined the maximum limit of paddlers permitted per day. The commercial operators determined the daily-use numbers that each company could allow.

With the new special-use permit system in place in 1983, the USFS recorded 61,000 people annually using the river. By 1994, that number had increased to more than 200,000 per year. That year, the USFS conducted a second carrying capacity study. From Memorial Day to Labor Day, 1,470 boaters completed USFS surveys at the Nantahala River take-out site. Collected data included the time of day and day of the week, the water release levels, the type of commercial and private-use watercraft and the daily number of people on the river (75 percent of the boaters were commercial users).

According to the study, increased water flow could potentially accommodate more users. In general, a one-hundred-cubic-foot increase in water flow could increase the commercial use of about 350 more people a day. Researchers recommended that the USFS, Nantahala River's recreational resource manager, work with multiple-interest groups to determine the most appropriate limits to protect the resource. "For dam-controlled rivers such as the Nantahala, results from carrying capacity and economic impact studies are essential for recreational managers to negotiate with power companies, outfitters, and other publics about the best use of impounded water for whitewater boating, hydropower and lake recreation."

NOC, the nation's largest canoeing and kayaking center, promotes "the use of recreational resources responsibly," but the number of whitewater users continues to grow. In 2016, NOC's 5,000,000th person rafted the Nantahala. The company often hosts whitewater training classes and sporting events. Balancing the growing numbers of paddlers while preserving Nantahala's natural beauty is an ongoing endeavor. To help provide the funds to protect the river's resources, the USFS collects about 3 percent of the receipts collected by the commercial outfitters. These funds have helped to maintain public launch sites, restroom facilities and picnic areas and to build safety railings and steps.

On Highway NC 1310, commercial outfitters launch watercraft on the Nantahala River near the power station of Nantahala Power and Light (purchased by Duke Energy in 1994). Near the juncture of NC 1310 and US 19, the USFS manages a public launch area at the mouth of Bowlin Creek. There, a crudely constructed chimney is all that remains of a 1939 house built by Utah Construction Company for Nantahala Power and Light employees. Across the river from the launch site, two chimneys mark the site of the Cole Fields. In about 1890, Matthew Cole's apple orchard had eight hundred trees. His grape vineyard, harvested by horse and wagon, produced nearly two thousand gallons of wine a year, a rare business in WNC during this period. Government agents stamped a seal of approval on each wooden barrel during inspections.

At the USFS launch area, independent paddlers pay one dollar a day or five dollars for a yearly pass. Water released daily from the dam at Nantahala Lake creates the eight-mile, three-hour whitewater adventure through Nantahala Gorge. About 75 percent of the route courses through NNF. Beginners and families navigate Class I–III rapids. At the public put-in, boaters navigate Nantahala's first major rapid, called Patton's Run, named for U.S. veteran Charlie Patton, a one-armed paddler who lost his limb during his military service. The river's corridor at Patton's Run turns almost ninety degrees, a geological footprint of significant natural events. Rafters follow Nantahala's big bend, a diverted riverbed carved centuries ago out of softer limestone pockets in its foundation. Nantahala River no longer travels north through Topton Gap to Robbinsville. Its turbulent forces have cut a new northeasterly route toward Fontana Lake.

"It's one of the most unique geological sites in the Southeast," said George Ellison in a Nantahala film created by him and Lance Holland. First explored and mapped by geologists in 1907, "the whitewater industry now generates revenue on events that happened maybe 100 million years ago."

Also of geological significance in Nantahala Gorge, the 190-acre Blowing Springs Special Use Area is protected by the USFS for scientific and environmental study. Resting on the Murphy Marble Geological Foundation, its geological characteristics make its soil content more alkaline, or sweeter, than the surrounding acidic soils. Unique plant communities grow there.

Since Nantahala River parallels US 19, motorists can watch paddlers navigate Patton's Run from a wooden observation platform built by the USFS. Nearby, a highway marker commemorates William Bartram's travels through this region in the 1770s. After paddling rapids past islands and large rocks, boaters pass under a bridge at Winding Stairs Road where the

Bartram Trail crosses the highway. Winding Stairs Road (FSR 422) curves past Queen's Falls and lake on a beautiful drive to Wayah Road.

Beyond waves, eddies and rapids, boaters pass an abandoned USGS water-level gauge before floating under the highway bridge near the Ferebee Memorial Picnic Area and Launch. (A new gauge records water levels about a quarter of a mile downstream.) After graduating from North Carolina State College, Percy B. Ferebee was a USFS engineer in 1914 working with surveyors mapping the Macon-Swain County line through the rugged Nantahala country. Soon, he left the Forest Service and became a banker, newspaper editor and mayor of Andrews. Throughout his career, he remained active in numerous civic organizations across the state. In 1920, he purchased the Nantahala Talc and Limestone Company in Nantahala Gorge. At his death in 1970, he left about five thousand acres of land bordering both sides of the Nantahala River to the people of the United States in memory of his wife, Florence, and their son, James B. Ferebee II. After the USFS acquired the land in 1973, the agency built the Ferebee Memorial Park. Members of the Older Americans Program helped the USFS build a comfort station, picnic area, observation platform and access ramp.

Downriver from the Ferebee Memorial, Nantahala Talc and Limestone Company is the oldest continuously operating quarry in North Carolina. Known for its exceptional specimens of marble, such as the pink-banded marble used in the construction of the Raleigh Courthouse, the quarry first opened in the late 1800s. In the early 1900s, the company mined talc. Today, its primary product is dolomite marble crushed into gravel. Geology students, science clubs, gem and mineral organizations and other rockhounds hammer out nice specimens of pink, yellow, purple and gray-banded marble, as well as soapstone, calcite, quartz, talc and aragonite. "At Nantahala Talc and Limestone Company," stated George Ellison in the film *Nantahala River*, "the geologic skeleton of the gorge is available for us to see today."

Rafters slip past whirlpools, slide down ledges, ripple over shallow shoals and glide under a swinging bridge. In years past, schoolchildren walked from home across swinging bridges like this one provided by the North Carolina Department of Transportation to catch the school bus. Boulders, eddies and one bend later, the whitewater adventure reaches its climax at Nantahala Falls.

Nearby, the USFS take-out area is located at the mouth of Silvermine Creek. In the 1980s, land exchanges between NOC and the USFS resulted in the NNF acquisition of the riverside property between Nantahala Falls

Nantahala River's lower section before it enters Fontana Lake. *National Forests of North Carolina Historic Photographs, D.H. Ramsey Library, Special Collections, University of North Carolina–Asheville.*

and the NOC restaurant. The Forest Service also acquired property along the A.T. corridor through the whitewater center. In the equal land value exchange, NOC received land near Silvermine Creek, the site of a lumber company's old log flume. When NOC built its footbridge across Nantahala

River between the store and restaurant, it incorporated the foundation column of the old flume into the support structure for its footbridge. Northbound A.T. hikers leave the Rufus Morgan Trail Shelter, hike through the Silvermine Creek Drainage area and cross US 19/74 and the Nantahala River on a footbridge to cross the heart of the Nantahala Outdoor Center Complex. Beyond NOC, A.T. hikers continue north toward the Cheoah Ranger District of NNF.

PART III

CHEOAH RANGER DISTRICT

With one stiff, downward beat of its wide, outstretched wings, a red-tailed hawk lifts off from a high perch in an old oak tree. Early morning sunbeams peek over the Snowbird Mountains, one of the most isolated, least-visited areas of North Carolina. A faint golden glow highlights the ridgeline, defining Graham County's southern border and the boundary between Cheoah and Tusquitee Ranger Districts. A light valley fog and mountaintop clouds slowly fade as the solar radiance casts its silent, magical spell. Coming into view, the lookout tower on 4,716-foot Joanna Bald rises above Tatham Gap Road. In the 1830s, army soldiers carved the old wagon road roughly following a former Indian pathway through the rugged mountains. Under federal orders, soldiers herded three hundred Cherokees out of their sacred Cheoah Valley to a stockade at Valleytown, now called Andrews. Cresting the Snowbirds was the highest point of their nine-hundred-mile walk to Oklahoma.

With grace and power, the hawk wings up higher. A wider view reveals tall, forested mountains rising steeply above the valley floor. Deep ravines glisten with the shiny silver of crystal-clear streams. Like children holding hands circling a playground, majestic mountains join shoulder to shoulder around the narrow, fertile valleys of Graham County, nearly encircling them. Unlike valleys spanning out into broad, open fields, however, the valleys of Cheoah District remain wrinkled with smaller, intersecting ridges. Dividing Tennessee's Cherokee National Forest from Nantahala National Forest, the Unicoi Mountains frame the area's western border. To the north, the Yellow

Notice the distinctive bellyband of the red-tailed hawk, often seen in flight. *Simon Thompson, http://www.birdventures.com/Calendar.html.*

Creek Mountains and the Cheoah Mountains highlight the Great Smokies rising in the distance. In the Snowbird Mountains to the south, the red-tailed hawk begins its journey over the last frontier settled by European pioneers in Western North Carolina.

Cruising along Nolton Ridge rising above Nantahala Gorge, the red-tailed hawk flies toward the grassy fields of 5,062-foot Cheoah Bald, where the yellow-blazed Bartram Trail greets backpackers hiking the A.T. from Georgia to Maine. The hawk spirals in mountain thermals above the ridgeline to Stecoah Gap. At Stecoah in May, serenades of spring migrants entice mates to the rich hardwood forests for nesting season. The endangered golden-winged warbler, known to nest in the brushy woodlands of Stecoah, will brighten some mornings.

Gently banking left toward the southeast, the raptor eases toward Robbinsville. Two centuries ago, its flight would have traversed a Cherokee town and Removal fort instead of the county seat. Here, in 1838, Fort Montgomery held Cherokee families removed from their homes by General Winfield Scott's army soldiers. Over the previous decades, white man's greed for gold and land had fueled one battle and deceitful treaty after another, robbing the Cherokees of their native homeland. At Fort

Montgomery and other forts, the Cherokees awaited their sad fate: the Nunahi-duna-dlo-hilu-I, Cherokee for "The Trail Where They Cried." In 1830, Congress had narrowly passed President Andrew Jackson's Indian Removal policy. By 1838, soldiers had begun the eviction of more than fourteen thousand Cherokees from forty thousand square miles in the Southeast and marched them west.

Although Junaluska's strategy and bravery in 1814 supposedly saved Andrew Jackson's life in the Battle of Horseshoe Bend, soldiers forced him to leave, too. Twenty-four years and selfish intent had erased Junaluska's daring feat, secretly swimming at night across the Tallapoosa River to steal the enemies' canoes. Walking nearly ten miles a day, Junaluska followed his people on the Trail of Tears. He and fifty others attempted an escape, but soldiers soon caught them. Nearly four thousand of his people died along the Trail.

Most settled in Indian Territory (present-day Oklahoma), but for some, it could never be home. They longed for the Appalachians, the land of their ancestors. When Junaluska walked back to Robbinsville, the North Carolina government awarded him state citizenship, a 364-acre farm and $100. Other Cherokees were also living in Graham County. Some had evaded the Removal by marrying white settlers. Others had hidden in the mountains. Some Cherokee people in other counties were able to hold on to land they had claimed in the Treaty of 1819, which the North Carolina courts upheld. Colonel William Thomas, the white advocate of the Cherokees, used his own funds and money earned by the Cherokee people to purchase land for them near places like Little Snowbird Creek in Graham County and the Qualla Boundary in Swain County. Now, both the Snowbird and the Qualla Boundary Cherokees live on land held by the U.S. government as trustee and designated as home for the Eastern Band of the Cherokee Nation. This is land that they bought back themselves after the Removal.

The Snowbird Cherokees considered Junaluska a leader, but he was never formally a chief. In 1858, he died en route to a Cherokee healing spring in Citico, Tennessee. In 1910, near downtown Robbinsville, oxen lugged a huge boulder up a steep, one-mile hillside to mark his grave. Embedded in the hard granite, a plaque provided by the Daughters of the American Revolution paid tribute to him and his wife, Nicie. At the dedication service, Reverend Armstrong Cornsilk, whose father knew Junaluska personally, paid tribute to his father's friend.

After Junaluska's death, W.W. Hayes bought his farm but eventually sold it. During W.B Ziegler's early 1880s journey through the Southern

Appalachians, he wrote, "The finest tract of land in the county is owned by General Smythe of Ohio and is called the Junaluska Farm. It is situated near the village on the banks of Long Creek and consists of 1500 acres, 400 or 500 of which are cleared valley land of rich, loamy soil." Later, George Walker owned the Junaluska property and, in 1911, sold it to Whiting Lumber Company. Junaluska's property is now part of Robbinsville.

The red-tailed raptor continues his regional survey. His perceptible wingtip primaries separate, fingering the slightest changes in air currents. He soars toward Massey Branch Ranger Station at Santeetlah Lake, where the convergence of Long, Tulula and Sweetwater Creeks forms the headwaters of the Cheoah River. In the 1930s, members of the CCC Massey Branch encampment helped the Forest Service restore forests, build campgrounds and picnic areas, form the trail system at Joyce Kilmer Memorial Forest and construct the buildings at the USFS Work Center. After the Paris Peace Accords were signed in the early 1970s, ending the Vietnam War, U.S. war supplies provided recycled materials to rebuild most of the Massey Branch Work Center buildings, including the gas station. According to retired USFS district ranger Scooter Brown, the agency hired staff through the Older Americans Program to complete the project. The senior citizen program in

Dr. John Spencer visits the burial site of the Cherokee leader Junaluska on a hill above Robbinsville at the Junaluska Memorial Site and Museum. *Photo by the author.*

Graham County, explained Brown, hired sixty to eighty older adults, one of the largest enrollments in the country at the time.

Maintaining a steady altitude, the hawk glides over Santeetlah Lake. Its dam impounds the waters of Cheoah River, East and West Buffalo Creeks, Little and Big Santeetlah Creeks and other tributaries. With its steep mountainous banks plunging underwater at the lake's edge, like those of an imaginary mountain-lake scene painted by the hands of romantic artists, Santeetlah Lake is one of North Carolina's most beautiful lakes. Fishermen catch bass and crappie, sunbathers catch rays on pontoon boats and green herons stalk prey on soggy snags. The hawk now turns west toward the Cherohala Skyway.

Sharing its roadway with Nantahala and Cherokee National Forests, the forty-mile-long Cherohala Skyway ribbons its way through high-elevation terrain between Robbinsville and Tellico Plains, Tennessee. Its name combines the first two syllables of one national forest with the last two of the other. In North Carolina, Cherohala offers trail access to Huckleberry Knob, Graham County's highest mountain and the site of a fatal plane crash. The high bald makes an ideal habitat for ground birds and small mammals. Rabbits, mice and voles entice hungry hawks and owls. Another hiking trail leads to Hooper Bald, crowned with gorgeous flame azaleas. Its open meadow once held a failed hunting lodge that introduced destructive, nonnative wild boars into the region. More trails lead to natural beauties and historic sites. Other visitors pack for picnics and cruise the Skyway in convertibles or on motorcycles.

At the Tennessee border along the crest of the Unicoi Mountains, the red-tailed hawk leaves the skyway and turns north. Soon, Joyce Kilmer–Slickrock Wilderness and its Tennessee neighbor, Citico Creek Wilderness, dominate the hawk's-eye view of this region. The USFS purchased the area from logging companies and now maintains it as a nationally designated wilderness. Protective measures prevent new roads and man-made structures. Trail maintenance projects require hand-held crosscut saws and blades, not mechanical ones. Joyce Kilmer Memorial Forest lies at the southern tip of the wilderness.

The northern tip of Joyce Kilmer–Slickrock Wilderness meets the banks of the Little Tennessee River below Cheoah Dam. There, Cheoah River empties its aquatic load, drained from the Yellow Creek Mountains, Santeetlah Lake, Slickrock Wilderness and other areas. The red-tailed hawk glides east over Cheoah Lake to Fontana Lake, where A.T. hikers cross the highest dam in the East. Looming on the northern skyline are the Smoky

Red-tailed hawk. *James Poling, Black Mountain, North Carolina.*

Mountain highlands, preserved as a national park. Shuckstack Lookout Tower stands as a historic salute to fire wardens of the past. Fontana Village Resort, located on the site of the town built in the 1940s for dam workers, invites visitors to stay, dine, hike or visit.

Farther east at Tsali Recreation Area, the hawk's broad wings, like large arms welcoming a warm embrace, gently cup invisible air currents and lower him to an open branch. On trail days designated for equestrians, horseback riders pass underneath the bare tree trunk to cross a shallow stream. On alternate days, mountain bikers will take this path. Do they know about Tsali, the Cherokee man who fled soldiers forcing him to leave his home? Do they see the red-tailed hawk? Do they know about "Tlanuwa," the mythical giant hawk of the Cherokee, that scans the Little Tennessee River for prey? Do they know that the hawk is considered by some American Indians to be an aerial companion to wind, rain and thunder to protect the creatures below him? After circling the Cheoah Ranger District, the red-tailed hawk perches above Tsali. Like a sentinel on guard at a military fort, the hawk rests above the historic site as raptor ancestors have done for generations before him.

JOYCE KILMER–SLICKROCK WILDERNESS

What does this taste like?" asked Dan Pittillo, a retired professor of plant ecology at Western Carolina University. He handed a slender, three-inch-long twig with its bark peeled back from one end to a member of the Southern Appalachian Plant Society.

"Wintergreen," she said.

"It's a black or sweet birch," he explained to the group. "It has three times more wintergreen, or methyl salicylate, than its relative the yellow birch. It used to be the primary source for oil of wintergreen."

In July 2015, Dr. Pittillo identified both species of birch along the figure-eight loop trail at Joyce Kilmer Memorial Forest. Easily recognized by its yellowish-brown peeling bark, the yellow birch tree grows in moist, well-drained slopes. So does its relative, the black birch, identified by its dark, smooth bark streaked with horizontal markings. Both birches provide beautiful wood for furniture, paneling and cabinets. When exposed to air, black birch wood becomes much darker, creating a less expensive alternative to tropical mahogany.

Although these two species of birch are more common in the forests of the Northeast, both grow at this altitude in the Southern Appalachians. Frequent woodland neighbors, also identified along the loop trail, included hemlock, maple, oak, yellow poplar and beech. A basswood shoots out several sprouts from the base of its trunk, increasing its chances for survival.

A huge Carolina silverbell, bending in a competitive search for light, hangs over the pathway. Its four-winged fruit distinguishes it from its two-winged Piedmont relative. Three types of magnolias grow in this lush forest, but the Fraser magnolia, with its twelve-inch-long ear-lobed leaves, is a Southern Appalachian endemic.

The smaller understory trees—like alternate-leaf dogwood, American hornbeam (ironwood), striped maple and witch-hazel—cast dappled shadows over thick wild gardens of ground-hugging plants. Wildflowers—some still blooming in July, others well spent—gathered on the forest floor. Tossed in the slight breeze, tall blue and black cohosh and scattered pockets of foamflowers hid the shorter Jack-in-the-pulpits and showy orchis. Ephemerals, like Dutchman's breeches, large-flowered trilliums, dwarf-crested iris and others, had already completed their annual performance. In March and April, these early risers had bloomed, pollinated and, possibly, reseeded before large trees had spread

DENDROLOGY 101

By Angela E. Brown, member services and technology coordinator, North Carolina Arboretum

At Western Carolina University, I walked into Dr. Dan Pittillo's class, knowing that pines had needles, oaks formed acorns and maples had leaves that looked like Canadian flags, and I didn't know a word of botanical Latin. On the first day, Dr. Pittillo began quizzing the class on the names of trees. I learned that many students, rattling off Latin names, had already completed dendrology at Haywood Community College. I had a sinking feeling that I was in trouble. We would get a grin from Dr. Pittillo, acknowledging that he understood that we knew the *common* name, but he would wait until we responded with the genus, species and family. No credit was given if you misspelled the name. Luckily, Latin is a phonic language, and I found it easy to learn and spell.

I loved our labs outdoors, rain or shine. Since I had a fall semester course, the summer-like weather during the first few months allowed us to wear shorts. Dr. Pittillo warned us several times that he didn't stay on the trails. We would be hiking through thick undergrowth, he would say, including poison ivy, but we still wore our shorts. (He their broad leaves, capturing the sun's energy for their own season of growth and reproduction.

With an annual rainfall around eighty inches a year and rich organic soils, this environment supports a wide variety of plant life. However, tiny, nonnative pests, called hemlock woolly adelgids, have invaded Southern Appalachian forests, including Joyce Kilmer, and have killed large hemlocks. To keep these giants from falling and injuring forest visitors, the USFS brought them down with dynamite, a controlled way to simulate straight-line wind damage. Without the dense canopy of hemlocks, increased sunlight and drier conditions have promoted the growth of other plant and animal species.

Alfred Joyce Kilmer died in France during World War I on July 30, 1918. The New York chapter of the Veterans of Foreign Wars proposed a living tribute to honor him. Officials in the national forest system found a suitable place. In a remote section of Graham County, quenched by the mountain waters of Little

quizzed us on its Latin name, of course: *Toxicodendron radicans*). One day, we took off straight up the hillside through poison ivy and weedy vines. Dr. Pittillo quickly lost all of us on the climb. At the top of the hill, we stood there breathless while he pointed out the dense growth of plants on our hillside scramble. When I returned to campus, I followed his advice, showered with Dawn detergent to remove the poison ivy oil and did not develop a rash.

The natural history lessons that Dr. Pittillo shared with us about the trees helped me remember the different species. The edible sourwood leaves (*Oxydendron arboreum*) have a sour taste. The flammable yellow birch bark is great kindling for a campfire, even if it's wet. I also learned the trees through word and shape association. For instance, since the leaf of the southern red oak (*Quercus falcata*) is bell-shaped at its base, I mentally equated it to a "southern belle." Its overall shape looks like a falcon's foot, thus the Latin "*falcata*." I collected leaves and made my own herbarium pages to study. I drew pictures of leaf scars and pine needle bundles. After my work and effort, I was excited with my grade, missing a "B" by three points. Dr. Pittillo was one of my most memorable professors. The class was hard, but I learned a lot. It boosted my confidence and helped me choose the college major that prepared me for my career.

"In the forest is one place I find my sanctuary," said Alexandra Holland, pictured in Joyce Kilmer Memorial Forest. *Steve Holland.*

Santeetlah Creek, massive trunks of yellow poplar trees lifted wide, green crowns one hundred feet toward heaven. This natural cathedral-like setting embodied the words of the sentimental poet: "A tree that looks at God all day; and lifts her leafy arms to pray."

For decades, Joyce Kilmer Forest has provided a field classroom in natural history for college students. Since the 1970s, Dan Pittillo has taught Western Carolina University students here. Retired USFS ranger William Nothstein brought Haywood Community College forestry students to Joyce Kilmer. Dedicated to Kilmer and his "Trees," a simple poem recited by schoolchildren for generations, a two-mile trail winds through the lush forest of Poplar Cove. What better woodlands could pay an everlasting tribute to a poet and inspire students than one of the greatest old-growth forests in the Southern Appalachians?

Since the late 1890s, several lumber companies have owned these forests along Little Santeetlah Creek, but none of them logged the current Joyce Kilmer area. According to John Bemis Veach, former president of Bemis Hardwood Lumber Company, the lumber companies of Bemis, Champion Fibre and Gennett collaborated in 1924 to purchase the timberland holdings of Whiting Lumber Company. The Gennett Lumber Company acquired the Little Santeetlah Basin. In 1929, it established a logging camp at the current Joyce Kilmer Picnic Area.

"We ought to buy it," wrote the Forest Service regional director in 1935, "to preserve some of the finest original growth in the Appalachians." When Gennett offered to sell the property at rates much higher than market value, the USFS agreed to pay its asking price. While other property owners were receiving three to four dollars per acre for lands purchased for the national forest system, the government, in 1936, paid Gennett Lumber

Schoolchildren take a field trip to Joyce Kilmer Memorial Forest in the 1940s—Doris Moody Ammons is at far left, and Max Moody is fourth from the left. *Lindy and Lisha Ammons.*

Company twenty-eight dollars per acre for its thirteen-thousand-acre Little Santeetlah tract.

On July 30, 1936, officials dedicated 3,800 acres of that tract as the Joyce Kilmer Memorial Forest. The poet's granddaughter attended the celebration. In his keynote address, the USFS assistant regional director read President Franklin D. Roosevelt's message: "[In] this living monument... his beloved memory is forever honored, and one of nature's masterpieces is set aside to be preserved for the enjoyment of [future] generations." The publicity attracted tourists from around the world. Before World War II, Ranger Nothstein guided thirty German foresters among the giant trees of Joyce Kilmer, mutually communicating in the universal scientific language of Latin.

In about 1936, the USFS also bought land from the Babcock Land and Timber Company in the adjoining Slickrock Creek watershed. Babcock had purchased the Slickrock area in about 1915 and heavily logged the area. When the Aluminum Company of America (Alcoa) built its 1930s hydroelectric dam, submerging Babcock's railroad under Calderwood Lake, the logging company ceased operations and sold the property. Babcock had previously sold its adjacent Tennessee tract at Citico Creek to the Forest Service. In 1975, the Citico Creek basin became a fifteen-thousand-acre wilderness study area in Cherokee National Forest. That year, in NNF, Congress joined the Slickrock and Little Santeetlah Basins, including the memorial forest, to meet the acreage requirements of the Wilderness Act of 1964. The act required that protected wildlands total at least ten thousand acres, but the memorial forest contained less than four thousand acres. Now, the protected combined area forms the Joyce Kilmer–Slickrock Wilderness. By 1984, the North Carolina Wilderness Act added about three thousand more acres.

John Denton and his family were the only recorded white settlers to have lived in today's Joyce Kilmer Memorial Forest. In the 1800s, Denton lived on land owned by Belding Lumber Company. For decades, the Cherokees had occupied this area. From their small settlement, located near the present USFS Rattler Ford Campground, the Cherokee families of Long, Ropetwister and Wachacha had traded goods along a well-known trading path through the Santeetlah area. In the 1700s, about four hundred Cherokees (half of the population) living here died from smallpox. Beneath tangled roots of tall trees and thick vegetation, they lie in unmarked graves.

Denton, born in Georgia in 1840, grew up in Polk County, Tennessee, as the son of a school commissioner. After serving in the Civil War, he married

Albertine Turner; the couple had nine children, five born in Tennessee and four born in Santeetlah's forests in North Carolina. In Benton, Tennessee, Denton worked as a jailor. His oldest child, Charles, was born in the jailhouse where they lived. In 1879, Denton moved his family across the Unicoi Mountains and into Graham County, North Carolina. Was he forced to leave after a dispute gone awry at a local dance, as some have claimed? Or did he, as a Confederate veteran of the Civil War, find it difficult to resettle in his Tennessee home, still recovering a decade after the brother-against-brother conflict? Was John embittered by his imprisonment at the fall of Vicksburg in July 1863, when he and others were pulled from caves dug into the banks of the Mississippi River?

If John Denton left his boyhood home resentful and bitter, he certainly would have found peace and inspiration in the stately safe haven of Little Santeetlah Basin. He chose his homesite near a giant, fallen chestnut log. After he carved out the dead heartwood, large enough for his tall frame to enter freely, John lived with his family in this "log home" while he built their cabin. Jane Meroney (Turner), John's mother-in-law and a cousin of Jefferson Davis, president of the Confederacy, was visiting Ohio relatives when the Dentons left Tennessee. She returned to her home in Benton heartbroken.

"She made her way alone and on foot to Graham County," wrote David Denton, great-grandson of John Denton and author of *The Denton Experience*. "When she sensed that she had found them and called like a bobwhite partridge from the banks of the Little Santeetlah Creek… Granddaddy John recognized the family signal and led her to the open arms of her daughter and grandchildren."

Years later, John Denton's eighty-year-old granddaughter, Oleta Rice Nelms, named for her mother's Cherokee childhood friend, Oleta Little Heart, worked as a USFS hostess at Joyce Kilmer Memorial Forest. In 1954, she retired from Robbinsville School after twenty-five years of teaching. With the Forest Service, Oleta worked with the Older Americans Program. Managed by the USFS and Department of Labor, the program aided senior citizens on fixed incomes.

"She was only paid minimum wage," recalled USFS Marshall McClung, "but she was worth a million dollars to the Forest Service." A 1987 newspaper article called her "a goodwill ambassador and hostess for the biggest 'cathedral' in the world." Mr. Kilmer would be proud, said the reporter: "Perhaps Joyce Kilmer would have called it poetic justice that the guardian of the forest dedicated in his honor is a former schoolteacher, considering that thousands of teachers have taught his famous poem, 'Trees,' to thousands more American children since its publication in 1914."

Ila Hatter, a former USFS naturalist at Joyce Kilmer, worked with Oleta in the 1980s and 1990s. Now a well-known specialist on edible and medicinal plants, Ila co-led the dendrology hike with Dan Pittillo that day in July 2015. "Thousands of Kilmer visitors from around the world loved to listen to her," said Ila. Once, after Japanese visitors returned home, enamored of her stories, Oleta was featured in a Japanese travel brochure.

Oleta was only a child when her grandparents John and Albertine Denton died in 1913. She had learned many stories from her mother, Mae, born at Santeetlah, and her mother's eight siblings. Oleta described the land cleared around the cabin by her grandfather for fruit orchards, gardens to raise corn and beans and pastures for horses, cattle, hogs and poultry. John and his oldest son, Charlie, who was eleven when they moved to Santeetlah, hunted for turkey and deer. In his article "Denton Family—Early Settlers," McClung, a Graham County historian, said that Charlie and his father "spent more time in the hunting camps in the mountains than they did at home…hunting fur-bearing animals and selling their pelts to local merchants." A ten-mile, Sunday morning walk led the family to Lone Oak Baptist Church, now submerged under Santeetlah Lake. On some Sundays, Charlie chose to "play with the Cherokee children instead."

In 1893, John Denton's sons, Charles and Forest, moved from Santeetlah to become caretakers of a one-thousand-acre farm in the Little Snowbird community. A few years later, John and the others followed. In a mid-1800s North Carolina land grant, Fredrick Augustus Waldo from Cincinnati, Ohio, had acquired the Little Snowbird property. After the land passed to his son, Frank Waldo of New Haven, Connecticut, in 1887, the Dentons became managers of the Waldo Cattle Ranch. According to Leota Denton Wilcox, curator of the Denton Family Museum and great-granddaughter of John, Waldo paid the Dentons for their service in parcels of land. Over the years, Forest Denton purchased more acreage from neighbors. When Graham County seized adjoining lots for non-payment of taxes, he purchased that property also. A tidy, well-tended Denton Museum now sits on the spot of an original Denton farm building destroyed years ago in a storm.

Words on a page and a family museum are all that remain to preserve Nantahala's history of the only white settlers in the Memorial Forest. Decay, leaf litter and a forest's annual growth now conceal a hardy mountain man's efforts to provide for his family. Only a weathered rock chimney remains, weakened by nature's elements and collapsing gradually with time. A poplar

Above: John Denton and his family, the only known white settlers to have lived in the area now known as the Joyce Kilmer Memorial Forest, lived in this cabin. *National Forests of North Carolina Historic Photographs, D.H. Ramsey Library, Special Collections, University of North Carolina–Asheville.*

Left: Retired Western Carolina University professor Dan Pittillo and naturalist Ila Hatter lead botanical hikes in Joyce Kilmer Memorial Forest. *Photo by author.*

tree sapling is growing atop the rubble, a tree often recognized as a pioneer itself, staking its own claim and taking root in old abandoned fields and cleared house sites. Perhaps nature itself planted one of the most regal trees of hardwood forests as its own memorial tribute to John Denton. Near the chimney, David Denton found "a frail lilac bush searching for sunlight under the canopy of the largest trees." On April 16, 2016, however, another transplant was found blooming—a Japanese rose (*Kerria japonica*), sometimes found at abandoned homesites.

The six-room Denton cabin with two chimneys is now gone, but it did not go easily. With long pegs carved out of black locust trees, wood known for its resistance to decay and shock-proof qualities, Denton built his poplar log cabin on a stone foundation to last a lifetime—or longer. Black cherry and white ash formed the floors. "So well constructed that in 1944, the USFS dismantled the Denton home place with dynamite," explained David Denton. "The logs were hand fitted, dovetailed, bored, and pegged with hardwood pegs, so the Forest Service workers, failing to dismantle Granddaddy John's handiwork by ordinary tools, resorted to high explosives." John Denton's mountain spring that supplied his family with cool water still breaks ground about seventy-five yards from the old cabin site, providing water for the Joyce Kilmer Picnic Area today.

Less than a mile from the Denton homesite, at the center of the Joyce Kilmer Loop Trail, a bronze plaque embedded in a large boulder honors a fallen poet. A one-hundred-foot-tall poplar tree offers a living tribute. On the guided July hike in 2015, Ila Hatter and Dan Pittillo stood beside its magnificent girth in front of the members of the Southern Appalachian Plant Society. Slowly, Ila picked up a green poplar leaf the size of a lumberman's broad hand.

"The Indians here in Santeetlah would pick up a large poplar leaf like this one to hold berries that they had collected," she said, with the leaf gracefully draping across her open palm. Ila folded the two side blades of the leaf toward the middle. Rolling the center blade inward, she enveloped the other two. Like a diaper pin, the leaf's stem woven into its soft green tissue held the berry-cup in place. Little Santeetlah Basin, part of the sacred homeland of the Cherokee people, is now a memorial forest of protected giants, abundant wildflowers and quiet pathways that provide moments of discovery, appreciation and reflection.

Today, volunteers with the Partners of Joyce Kilmer–Slickrock Wilderness join civic groups and the USFS to help maintain the area. On April 29, 2016, twenty-six Partner volunteers helped clear brush, erected a new entrance sign,

SPRING WILDFLOWERS IN JOYCE KILMER MEMORIAL FOREST

By Scott Dean, field botanist/educator at the North Carolina Arboretum and http://westernncnaturally.com

The Memorial Forest is a remnant of an old-growth forest, containing more than one hundred species of trees, with a few about four hundred years old. On one walk, I counted forty-three species of wildflowers. Some of the beauties seen April 9–12, 2014, included the following.

Upon entering the loop trail, one of the first flowers you will see is dwarf ginseng (*Panax trifolius*), with a whorl of three lobed leaves. It shares many of the medicinal properties of American ginseng, but due to its small size and relative scarcity, it has never been used extensively in medicinal applications.

Rue anemone (*Anemonella thalictroides*) has two to three fragile, stalked flowers above a whorl of three basal, lobed compound leaves. The leaves grow in three groups of three on long stalks. Indians used a root tea for diarrhea and vomiting.

Wood anemone (*Anemone quinquefolia*) has no petals but five regular white sepals that look like petals. The foliage is deeply cut, often into three to five leaflets. A northern species, it is a relic of the last glaciation, some fourteen thousand years ago and has adapted to the mountains.

Sharp-lobed hepaticas (*Anemone acutiloba*) were thought to be a treatment for liver ailments because of the three lobes of the leaf. This was such a widespread concept that, in 1883, some 450,000 pounds of the dried leaves were commercially processed.

Usually found at higher elevations, squirrel corn (*Dicentra canadensis*) has a complex flower that restricts its pollination to insects, usually bumblebees, who hang from the bottom of the flower and vibrate their thorax, loosening and collecting the pollen as it falls through the opening at the bottom of the flower. The root's resemblance to kernels of corn led to the common name. As is common with *Dicentra* species, this is poisonous to cattle.

You will see, in great abundance, wake robin or red trillium (*Trillium erectum*), a member of the Lily family (*Liliaceae*). American Indians used root tea to treat menstrual disorders and to induce childbirth. This practice was shared with the early settlers and led to the common name "Bethwort" from the term birthwort. This species may have a red or white flower and has also been used to ease menopausal symptoms, as well as an aphrodisiac. It is sometimes referred to as "stink robin" due to a somewhat unpleasant aroma. Typical of all trillium, it blooms from April through June. Both color morphs occur here.

upgraded the handicapped-accessible picnic area and removed nonnative invasive plants at the Memorial Forest. Each year, the Partners offer a synchronized firefly walk at Joyce Kilmer, host local events, organize festivals and lead public hiking trips. In cooperation with Tri-County Community College and the USFS, the Partners also sponsor a Junior Forestry Summer Camp for rising sixth to twelfth graders. In 2015, day campers learned about search-and-rescue, trail management, forestry, dendrology and ecosystems.

"Firefighters showed us how fire burns a forest in different ways," said twelve-year-old Joey Muehlhausen. "They showed us how they created fire breaks and how wind affects a fire." When the young participants put on Personal Protective Equipment, "It was heavy," said Joey, "and hot even without the fire!"

Joey's family moved to Graham County for "a good stable foundation in the natural environment," said Lisa Russo, Joey's mother. "Hiking is my passion. I wanted them to enjoy it and they do. Many kids today have nature-deficit disorder. Through 4-H, Boy Scouts and programs offered by the USFS and the Partners of Joyce Kilmer–Slickrock Wilderness, Joey and his sister, Lili, can learn about our forests through outdoor recreation."

Joyce Kilmer Memorial Forest forms the southern edge of Joyce Kilmer–Slickrock Wilderness. The wide southern expanse of the protected wilderness narrows to an arrowhead-like point at its northern border with the Little Tennessee River. US 129 snakes nearby. Beneath the highway bridge, Calderwood Lake engulfs the Little Tennessee River and squeezes west through a narrow mountain valley framed by high rocky bluffs. To the east, the 225-foot Cheoah Dam, near the mouth of Cheoah River, dominates the view. In 1916, Alcoa built an extension of the Southern Railway line from Knoxville to its Calderwood community in Blount County, Tennessee. Later that year, the company extended that rail line to its dam construction site at the confluence of the Little Tennessee and Cheoah Rivers to transport supplies and workers. Completed in 1919, the Cheoah Dam was the first of several Alcoa hydroelectric dams built to supply power to its aluminum smelting company near Maryville, Tennessee. During the war years, power produced at Cheoah helped meet the increased demands for aluminum. At the time, it was the highest overflow dam in the world using the world's largest turbines. It also boasted the world's longest transmission lines (5,010 feet) with the highest amount of voltage (150,000 volts). Later owned and maintained by Tapoco, a subsidiary of Alcoa, the dam is now listed in the National Register of Historic Places.

In North Carolina, several development and power companies combined to form the Tallassee Power Company to manage the power station at Cheoah Dam. The company held a contest for a new name for its village. The twenty-five-dollar winner combined the first two letters in the company's three-word name to create Tapoco. From 1916 to 1919, along the banks of the Cheoah River, at its confluence with Yellowhammer Branch and Meadow Branch, the Tapoco community housed nearly two thousand dam and power station families. The company provided bunkhouses, individual homes and dining halls. It built a hospital, school, theater and country store. In 1920, a new railroad bridge crossed the Little Tennessee at Cheoah Dam, replacing the one destroyed by a flood. Until a road was built to Tapoco in the 1930s, boats on the river and a rail bus on the railroad provided transportation into the area.

In 1926, while building its dam to form Santeetlah Lake, Alcoa extended the Cheoah Dam railway upstream along Cheoah River. After completing Santeetlah Dam, the company discontinued the rail line. Later, it transferred its old road into the area to the NC Highway Commission. The new state road eventually became US 129. In 1931, the commission completed the US 129 route to Deals Gap, connecting the region to Bryson City, Maryville and Knoxville, Tennessee. This section of roadway, the "Tail of the Dragon," twists into 318 curves over eleven miles, a motorcyclist's dream.

The Tapoco Lodge, built by Alcoa for its company executives, later opened as a tourist resort. Owners added a gas station, tennis courts, a theater and a swimming pool. Listed in the National Register of Historic Places in 2004, the Tapoco Lodge offers visitors dining services, lodge suites and private cabins. The restored historic theater, Tapoco Tin Movie House, once hosted a "bon voyage" party for World War II soldiers. Nearby trails into NNF lure guests to some of the best swimming holes in the state. During the periodic reservoir release at Santeetlah Dam, the Cheoah River from Robbinsville to Tapoco Lodge becomes a Class IV and V whitewater playground for experienced kayakers.

Near Tapoco, one of the most scenic trails into the wilderness is accessible from its trailhead near the bridge at Cheoah Dam, where actor Harrison Ford escaped a U.S. marshal (Tommy Lee Jones) in the 1993 film *The Fugitive*. After a quiet walk along Calderwood Lake beside yellowwood trees and dwarf rhododendron, the Slickrock Creek Trail follows Slickrock Creek, roughly defining the North Carolina–Tennessee state line. In April, blankets of spring wildflowers slow a hiker's pace, like devotees of art in a fine gallery. Bishop's Cap, foamflower, bellwort, chickweed and a variety

"MUST SEE" PLACES ON THE CHEOAH RANGER DISTRICT

By Hoot Gibbs, retired USFS Technician, now volunteer; search-and-rescue personnel

The signed parking area for Yellow Creek Falls is located about four miles south of Tapoco Lodge on NC 129. An easy, one-way quarter-mile trail leads to the beautiful eight- to ten-foot-high waterfall.

Wildcat Falls and Lower Falls require several miles of hiking but offer great scenery and swimming. The best access is from Big Fat Gap Trailhead at the end of FSR 62. Consider arranging a shuttle to Big Fat Gap and hike down Slickrock Creek to either the Tapoco Trailhead or the Tapoco Lodge via the Yellowhammer Gap Trail.

Many consider Middle Falls the prettiest on the Cheoah District. From the parking area at the "Junction" at the end of FSR 75, follow Trail No. 64 beside Snowbird Creek on an old railroad bed. At three miles, pass Sassafras Creek. Take Trail No. 65 for about one mile to Sassafras Falls. Continue along Snowbird Creek for another two miles to reach Middle Falls.

Cheoah Bald, with great scenic views, is best visited during October, when Nolton Ridge Road (FSR 259) is open for the fall color season. Park at the end of the graveled forest service road. Hike north along a woods road and trail for a moderately strenuous three-mile round-trip hike.

The Hangover, a rock cliff overhang, offers a 360-degree view. The strenuous hike from Wolf Laurel Trailhead is approximately six miles round trip. The fall color season is absolutely gorgeous! Friends and I hike to Hangover each fall and are blessed to be in the midst of God's beautiful creation!

of trilliums bring photographers to their knees, put pens to pads and uplift solemn spirits. Cold mountain streams tempt fishermen. On lucky days, the Junaluska salamander, a species that lives nowhere else in the world, will make an appearance. Thirteen-mile loops beckon hardy hikers into the wilderness. Three-mile hikes bring others to destinations, like the white curtain of Lower Falls breaking the surface of a waiting pool.

On some trails, rotting railroad ties interrupt a hiker's steady stride. In the early 1900s, Kitchen Lumber Company and, later, Babcock Land and Timber Company logged this area. By extending Alcoa's rail line from

Cheoah Dam into that area, lumbermen moved logs to Tennessee markets by rail instead of rafting them on the Little Tennessee River, as smaller companies had done. South of Tapoco and east of Nantahala's wilderness boundary, spur railroad lines followed Deep, Barker and Bear Creeks. When Alcoa constructed Calderwood Dam, lake waters submerged the railroads and logging in Slickrock ended.

Experienced hikers often enter Slickrock Wilderness at Big Fat Gap to enjoy distant views, walk along razor-like edges or get high on Hangover. The Naked Ground Trail, named "Pleasant Gardens" on an early 1900s map, follows an old Cherokee trading path. A green forest now clothes Naked Ground, no longer bare like it was when the Cherokees named it. A good map, compass and GPS are recommended for exploring the wilderness areas. Invite a companion. Let friends know the planned route and destination. Prepare for the unexpected and be aware of the weather.

Backcountry hikers find solitude in Slickrock Wilderness. However, its rugged challenges can lead to tragic events. In the fall of 1988, a dramatic search began. On September 25, twenty-five-year-old Jim Michelic from Wisconsin purchased a map at the Cheoah Ranger Station. He wanted to "get away from people" for a while. After Jim spent the night at a hotel in Andrews, hikers saw him the next day on the Wolf Laurel Trail. When Jim had not returned on October 3 to meet friends in Atlanta as planned, they reported him missing. The USFS launched a search-and-rescue mission. Two hundred federal, state and local officials joined search dogs and local volunteers to comb the area. Women of the community provided food and support.

At the Wolf Laurel trailhead, searchers found Jim's car with its Wisconsin license tag. Others searched Stratton Bald, Naked Ground, Hangover and other trails. Heavy rains and dense fog hampered their efforts. Thick foliage reduced visibility for aerial surveillance. On October 4, Hoot Gibbs and Marshall McClung, both retired USFS foresters and experienced search-and-rescue volunteers, found an unoccupied tent site on Stratton Bald. High in a tree, the camper had tied food beyond the reach of prowling wildlife. "The terrain was steep and rugged," said McClung, "with underbrush so dense that crawling on all fours was necessary to get through." Finding no signs of Michelic, the organized search intensified, to no avail.

On October 16, two fishermen bushwhacked through rhododendron to the headwaters of Little Santeetlah Creek. At the base of a thirty-foot cliff, they found boots and a backpack. Michelic's brothers identified his personal belongings. The next day, the fishermen led rescuers off Wolf Laurel Trail into the rocky ravine. There they found Michelic's body and other items. An autopsy revealed severe neck and head injuries, presumably from a fatal fall. Restricted by heavy undergrowth in his perimeter search of Michelic's camp, McClung later realized that "we came within one-tenth of a mile of the victim on the first day." At the Stratton Bald campsite, the Michelic family returned to spread Jim's ashes.

In the Cheoah District, most search-and-rescue operations occur in Joyce Kilmer–Slickrock Wilderness. Not all victims are backcountry hikers. There have been reports of lost children, older adults and several members of a group. Some rescue missions require a search for survivors in downed aircraft. In August 1958, a Piper Tri-Pacer flew from Detroit to Knoxville. At 7:30 p.m., the plane departed Knoxville Airport en route to Atlanta, but it never arrived. Civil Air Patrol units from Tennessee, North Carolina and Georgia searched a broad area without success. At Fontana Village, a gas station attendant reported seeing a low-flying aircraft. At Calderwood, residents saw a small, low-cruising plane headed southwest toward Calderwood Dam in front of a brewing thunderstorm. The pilot, they thought, seemed to be following the Little Tennessee River as a navigational landmark. Initially, strong storms and low cloud ceilings limited aerial searches, but six days later, a small-aircraft pilot spotted the Tri-Pacer, crashed about a mile south of the Haoe Lookout Tower.

At the time of the crash, an abandoned fire tower sat on top of Haoe Lead, the east–west ridge that stretches from Naked Ground to what is now known as the Maple Springs Overlook. This rocky ridge separates the Joyce Kilmer Wilderness area from the wilderness section of Slickrock Creek. (The USFS removed the fire tower on 4,600-foot Haoe Mountain in the 1970s when the area was designated as a nationally protected wilderness.) Rescuers hiked to the site to recover the bodies of two men. The pilot's broken watch had stopped working at 7:50 p.m. Investigative reports concluded that the pilot experienced turbulent weather as he approached Haoe Lead. In an attempt to clear the ridge, the pilot's full-throttled, nose-up configuration stalled the small aircraft, spiraling it into a dive. The plane's left wing clipped the trees before it plunged nose-first into the ground.

In 1998, another small aircraft and its pilot, the father of FBI agent Jerome Barker, also went missing. Jerome Barker and other FBI agents had come to NNF to search for Eric Rudolph, the fugitive avoiding arrest for anti-abortion bombings. The agency established its headquarters in Andrews, North Carolina. When Barker's father, Walter, decided to fly a

Piper Arrow from Ohio to Andrews for his son's birthday, he never made it. Since Walter had not filed a flight plan or contacted any air traffic controllers while en route, an extensive search ensued but never revealed any clues. For six years, Jerome returned to hike the Nantahala region looking for his father's plane. In October 2004, a bear hunter in the Deep Creek area of the Kilmer–Slickrock Wilderness told District Ranger Joe Bonnette that he had discovered the site of a plane crash. The aircraft identification number matched Barker's. Scattered for 275 feet at an elevation of 3,300 feet, the wreckage pile indicated that Barker had been flying well below local mountain heights.

CHEROHALA SKYWAY

Beside Long Creek in Graham County, U.S. soldiers expanded an existing trail for the forced march of the Cherokees through the Snowbird Mountains. Cherokee families detained at Fort Montgomery, built on a Cherokee village ballfield, marched over the Snowbirds to Fort Delaney in Valleytown and remained there until soldiers moved them west to Fort Butler in present-day Murphy. (Today, FSR 423 retraces a portion of that section of the Trail of Tears.) Lieutenant James Tatham and his son surveyed that route, making it Graham County's first wagon road. Although the Tatham Gap Road was rough and rugged, it made this remote region more accessible.

The lack of good roads limited the region's population and economic growth. In 1872, Cherokee County relinquished lands to form Graham County, named for North Carolina governor (and later U.S. senator) William A. Graham. The new county's post office, formerly called Cheoah Valley and then Fort Montgomery, became Robbinsville. During Judge David Schenck's visit to Robbinsville on his 1880 Western Circuit tour, he left his overnight stay with Dr. Washburn at the headwaters of the Valley River, bound for the county seat. "Robbinsville was a hamlet of sixty-one persons," he wrote. "The entire county population is 2335....Roads were poor and most of them lying at a 30% angle."

Rough roads and footpaths followed old wildlife foraging trails. Suspension bridges spanned Little Snowbird Creek. Covered bridges crossed others. Logging companies, arriving at the turn of the century, improved and built many roadways, like Bemis Lumber Company did in Robbinsville. In 1902, a new road from Topton to Robbinsville provided lumbermen access to the Southern Railway's Murphy Branch. Twenty years later, convict labor widened Topton Road (US 129). By 1931, US 129 extended west of Robbinsville, past Santeetlah Lake and on to Tapoco at the Tennessee state line. In the 1950s, the improved US 129 began attracting tourists. In a 1961 edition of *The State*, Bill Sharpe wrote:

> *Graham was the last land to be wrested from the Indians and settled by white men....* [F]*or a trip of absorbing beauty and interest, plan a trip to Graham County, North Carolina's last frontier....Graham's perpendicular geography is hard to duplicate. The space between* [the mountains] *is not, however, either valley or plateau, but a maze of ridges interspersed with stream-cut valleys, some of them deep enough to rate as canyons, like the gorge of the Cheoah River below Lake Santeetlah. Recent improvements of US 129 from Topton has cut travel time to Robbinsville in half. NC 28 from Fontana Dam through Stecoah opened up the northern end of the county.*

With that same progressive spirit, members of the Tellico Plains (Tennessee) Kiwanis Club suggested a scenic route across the Unicoi Mountains through Nantahala and Cherokee National Forests, connecting their town to Robbinsville. To garner support and attract attention, club members arranged an annual, old-fashioned wagon train across USFS roads and trails. That first year, sixty-seven covered wagons and three hundred horseback riders joined the procession. To boost local economies and encourage tourism, politicians promoted the "Robbinsville-Tellico Plains Highway" (or the "Overhill Byway," as some USFS officials called it). Proponents of the project believed that the scenic route, now known as the Cherohala Skyway, could rival the Blue Ridge Parkway or the Great Smoky Mountains National Park's Newfound Gap Road.

In 1962, Congress approved the plan. Since the route would traverse two national forests and Congress had passed the 1965 Appalachian Regional Development Act, granting funds to improve southern highways, the U.S. government paid for the construction. By 1970, Tennessee had completed most of its twenty-one skyway miles. Almost immediately, North Carolina began construction, too, but stopped at the Maple Springs Overlook. Rising costs and budget cuts delayed the project—so did environmental voices raising other concerns.

Wagon trains promoted the construction of a scenic highway from Tellico Plains, Tennessee, to Western North Carolina. This one arrives in downtown Murphy. *Cherokee County Historical Museum.*

Across the country, local governments promoted tourism to improve local economies. Some questioned what impact such projects would have on the natural environment. In January 1970, the government signed into law the National Environmental Protection Act (NEPA). The NEPA mandated that any proposed project funded by the federal government must be evaluated by biologists, geologists or others for its potential impact on the environment. The environmental concern of the Cherohala's construction across the Unicoi highlands fueled years of protests, petitions, debates and delays.

Carl A. Reiche, an accountant in Coral Gables, Florida, vacationed frequently in Western North Carolina. As president of the Save-Joyce-Kilmer League, Reiche claimed that the planned roadway was "one of the worst political and bureaucratic examples of skullduggery ever known." He entitled Appendix A of his memoir "The True and Complete Story Behind the Threatened Environmental Destruction of One of the Most Impressive Remnants of Our Nation's Virgin Wilderness." He claimed that his documents exposed "the official web of falsehoods." More than twenty

North Carolina and Tennessee conservation groups joined the campaign. Their primary concern was that the proposed route across Haoe Lead and the ridge dividing Little Santeetlah and Slickrock watersheds would degrade both headwater basins. Road construction along that route would also impact the protection of Joyce Kilmer Memorial Forest.

"I called Congressman Roy Taylor," said Dan Pittillo, a regional conservationist with a University of Tennessee–Knoxville group. "I told him that if they built the Robbinsville-Tellico Plains Road in that location, it would prevent that area to ever qualify for protection under the Wilderness Act of '64. For land to be included in the wilderness system, it had to total ten thousand areas. Their road would have prevented that."

"Roads are made by fools like thee; but only God can make a tree," chanted protesters at Joyce Kilmer Memorial Forest on a cold, drizzly October day in 1970. Traveling from five southeastern states, eighty environmentalists gathered on Kilmer's trails. One had hiked more than seven miles to attend. Ted Snyder, president of the Joseph Le Conte chapter of the Sierra Club, laid a wreath on the giant boulder that held the bronze plaque in memory of Joyce Kilmer and his "Trees." To the tune of "I've Been Working on the Railroad," some sang, "We've been walking through Joyce Kilmer all the livelong day; we've been walking through Joyce Kilmer just to keep the road away!"

The *Christian Science Monitor*, the *Atlanta Journal*, the *Charlotte Observer*, the *Chicago Tribune*, the *New York Times* and others covered the event. Headlines like "Highway to Bisect Forest," "Road Cited to Threat Wilderness" and "Will Kilmer's Trees Survive?" attracted national attention. The local *Graham Star* reported, "New Road Would Destroy Environment: Rape of the Virgin." At a time when the nation's awakening conscience embraced the need to protect irreplaceable wilderness areas, voices rose in protest over the new Cherohala highway. The public wrote letters, attended meetings and signed petitions. According to Reich, the "increasing storm of public opposition stopped the bulldozers in their tracks."

In 1975, when Congress protected the Joyce Kilmer–Slickrock region as a wilderness area, the designated route of the thirty-six-mile Cherohala Skyway became official. Fifteen miles would cross North Carolina. The additional mileage of entrance roads at each end now totals fifty scenic miles from Robbinsville to Tellico Plains. Opened in October 1996, more than thirty years after Congress approved its construction, the $100 million roadway is now designated a National Scenic Byway. In 1971, Mary P. Stephenson wrote in *Wildlife in North Carolina*:

> *There may well be a need for improved roads in this part of Western North Carolina and eastern Tennessee, but presently there are several fine scenic roads in the Nantahalas—the Wayah Crest Road, the Nantahala Gorge and Highway 129 from Topton to Tapoco. Proponents and opponents of the road would do well to consider that Kilmer Forest is unique and one of the very last areas of wild country in the East.... The trees of Joyce Kilmer cannot speak for themselves; it is left to mankind to* [be a] *partner of all living things, not plunderer.*

West of Santeetlah Lake, near Joyce Kilmer Memorial Forest, the Cherohala Skyway begins at Santeetlah Gap. North of here, near the USFS Horse Cove Campground, the remnants of the four-mile controversial skyway route lead to the handicapped-accessible Maple Springs Overlook. At the skyway entrance, the Old Wagon Road (FSR 81) veers off to the right and parallels the southern border of Joyce Kilmer–Slickrock Wilderness. Also called Stewart Cabin Road, the graveled twelve-mile route follows Big Santeetlah Creek, known for its great trout fishing. (Staff at the ranger station can provide current road conditions.)

In less than four miles, a cabin built by James Archibald Stewart and his wife, Catherine, transports visitors to life in the 1870s. Before a flood in 1895 diverted the creek and destroyed a corn mill, the cabin sat one-half mile downstream on the banks of Big Santeetlah. The USFS moved it to a grassy knoll, shaded by oak, walnut and poplar trees, and now maintains it as a National Historic Site. A retired USFS fire warden for the Haoe Lead Lookout Tower, Sam Adams, helped restore the cabin. As a skilled carpenter with the Older Americans Program, he constructed its reroofing materials. An open-weave springhouse sits out back.

Stewart Cabin Road passes picnic areas and campsites. Dead-end spur roads offer backcountry adventures. Dainty maidenhair ferns soften roadway banks. Tall stems of black cohosh wave petal-less flowers, formed by tightly clustered stamens around a central stigma. In July, beebalm, coneflower, black-eyed Susan and other wildflowers enhance the tour. Bumblebees bump and bully other bees for nectar from purple-flowering raspberry. Pipevine swallowtail butterflies dance around a fire pit, alighting for a few seconds on charred wood chips and then twisting again, more interested in their fluttering carefree capers than a visitor's admiring attention. Red-bellied woodpeckers drum a dead pine, while a red-tailed hawk screeches his downward vocal spiral overhead. The background beat of Big Santeetlah's tumbling waters drown out fainter, audible songs of the forest.

At nine miles on FSR 81, a right turn onto Wolf Laurel Road leads to the primitive, three-room Swan Cabin, built in 1931 by Frank Swan. The Forest Service later moved it to this site. A per-night reservation fee rents the rustic accommodations. Guests roll out sleeping bags on nine rope-swing bed frames, draw water from a spring, warm up by a wood heater, take in sweeping views from Swan Meadows and hike to Stratton Bald or nearby waterfalls. FSR 81 joins Cherohala Skyway at Stratton Meadows about a mile from the North Carolina–Tennessee border. A turn toward the east returns drivers along the skyway to Santeetlah Gap and Robbinsville; a westward choice leaves NNF and enters Cherokee National Forest en route to Tellico Plains.

At the Santeetlah Gap entrance to the Cherohala Skyway, exhibits provide valuable information and suggest to visitors that they take their time along the two-hour cruise. Overlooks provide long-range vistas and picnic tables. At picnic areas, bear-proof trash containers discourage easy foraging but encourage humans to "Leave no trace." Members of the Natural Resources, Recreation and Water Quality Committee for the Graham Revitalization

The NCWRC erected three pairs of utility poles to promote the safe gliding of Carolina northern flying squirrels across the highway. *Christine Kelly, NCWRC biologist.*

Economic Action Team (GREAT) volunteer to help the USFS and other groups maintain the skyway. In the spring of 2016, they collected forty bags of trash, mostly aluminum cans, and cleared brushy overgrowth from scenic overlooks. Annually, the group bags more than two hundred sacks of trash from local roadsides and bicycle trails.

Birds and wildlife flourish along the skyway. Nocturnal wild boars normally roam the backcountry, but white-tailed deer and black bears may cross Cherohala. To help federally endangered Carolina northern flying squirrels (CNFS) safely glide across the highway between denning and foraging sites, the North Carolina Wildlife Resources Commission (NCWRC) erected three pairs of modified utility poles in 2008. From 2009 to 2010, four wildlife cameras captured fifteen still images and twenty-five videos of CNFSs exploring the structures. About 56 percent of the flying squirrels leaped across the road from beam to beam. The NCWRC radio-tracked four individuals and also captured and tagged others at nest boxes. Biologists later recaptured some CNFSs on the opposite side of the road, reconfirming the success of the utility poles.

"We believe," said Chris Kelly, NCWRC biologist, "that these structure-assisted road crossings in the Unicois are the first observed in a North American gliding mammal."

In 2011, the NCWRC captured a flying squirrel near the Cherohala at 4,073 feet in a cold, north-facing drainage, the lowest recorded denning site to date. Although the CNFS is related to the common southern flying squirrel, it normally inhabits cold, coniferous forests in the northern United States and Canada. Rare in the Southern Appalachians, this northern species only lives in nine separate high-elevation forests. Fragmentation of habitat is a significant concern.

At Whigg Cove on the Cherohala Skyway, the NCWRC and USFS treated hemlocks dying from hemlock woolly adelgids, but the big coniferous CNFS homes did not respond well. In 2013, the agencies began a habitat improvement project by planting red spruce to replace the dying hemlocks.

"Red spruce probably hasn't occurred naturally in the Unicois for thousands of years," said Kelly. "Thus, this was not a true 'restoration' project, more a conservation measure to maintain the conifer component for the CNFS."

NCWRC collected red spruce cones in the Smokies and the Great Balsams. The Southern Highlands Reserve in Transylvania County, North Carolina, extracted the seeds and grew seedlings. In 2013, at Whigg Cove on Cherohala Skyway, the USFS, NCWRC, U.S. Fish and Wildlife Service and Haywood Community College forestry students planted about 1,100 seedlings. In 2015, Warren Wilson College students joined others from Haywood Community College to help plant another 500. Biologists monitor the plots each fall. In 2016, the young spruce seedlings were healthy. The USFS and NCWRC began "release work" to provide greater sunlight to promote faster growth.

Driving west on Cherohala Skyway, motorists find great views into the Santeetlah Basin from the Hooper Cove (3,100 feet) and the Obediah (3,740 feet) Overlooks. Stratton Bald rises to the west. Haoe Bald, and its ridgeline, Haoe Lead, take center stage. Horse Cove Ridge, the nearest, most prominent divide, separates Joyce Kilmer Wilderness from the Slickrock Wilderness area. Nearly forty years after the Forced Removal, Chief John Obediah of the Cherokee town of Cheoah voluntarily led about ninety Cherokees out of Graham County to join relatives in the West. Receiving help from the Indian Agency in Loudon, Tennessee, they boarded a train for the Indian Territory out west.

Consider walking a nature trail through a hardwood forest at Wright Cove Overlook. At the Spirit Ridge Overlook (4,950 feet), a handicapped-accessible trail courses through an open forest with exhibits describing the natural history of beech, yellow birch, buckeye, cherry and other trees. As if planted by the University of Tennessee football fans, Turk's-cap lilies gently wave their orange flags. At the end of the quarter-mile trail, a wooden observation platform overlooks Cherohala Skyway.

At the Huckleberry Knob parking area, a two-and-a-half-mile round-trip hike on an old roadbed leads to Cheoah Ranger District's highest summit. Huckleberry Knob (5,560 feet) offers wide-ranging views of the Unicoi and Great Smoky Mountains. In the summer, indigo buntings burst with color and song from open treetops, while eastern towhees scratch under brushy vegetation. By mowing the knob once a year, the USFS and NCWRC maintain Huckleberry as foraging field for wildlife.

Huckleberry Knob and other mile-high mountains along the skyway endure severe wintertime weather. Sometimes the extreme elements lead to tragic outcomes. On a snowy day in December 1899, two Heiser Lumber Company employees left Tellico Plains on foot, bound for Robbinsville. They never arrived. In September 1900, Forest Denton, avid hunter and son of the Joyce Kilmer settler John Denton, found the remains of Andy Sherman and Paul O'Neil, a whiskey jug and an old campfire site. Today, a gravestone and white cross mark Sherman's grave on Huckleberry Knob. In the early 1900s, the Graham County court system donated O'Neil's intact skeleton to Dr. Robert J. Orr of Robbinsville for a medical exhibit.

In 1952, Vick Denton, another son of John Denton, worked as a game warden for the NCWRC. When an emergency call reported a plane crash on Huckleberry Knob, Denton, USFS officials and others responded. After departing a Nashville airfield, military pilots in two F-51 fighter planes had encountered low clouds and severe turbulence. One pilot had turned back. The other, thirty-one-year-old Captain Donald H. Meode, had perished when he plowed into the side of Huckleberry Knob. Rescuers found his body and carried it down the steep terrain to a waiting helicopter on Hooper Bald.

In 1989, the Middle Atlantic Chapter (MAC) of the American Rhododendron Society studied the native azaleas on several balds in the Southern Appalachians, such as Gregory Bald in the Smokies and Hooper Bald on the Cherohala Skyway. The growth of brambles and other vegetation was encroaching on these native beauties. In 2004, MAC met with NNF officials to propose a "Hooper Bald Project," clearing competitive vegetation from around each plant, as well as collecting and planting azalea seeds. Since then, MAC and other volunteers have tended the native azalea garden each year. The USFS does not permit selling the azalea seeds but promotes the collection and distribution of the seeds to perpetuate these specialties. In addition to the "Hooper Copper," whose three-and-a-half-inch-wide golden flowers turn a deep orange with age, five other Hooper Bald azalea species grow nowhere else in the world.

On Cherohala, just west of Whigg Cove, named for early settlers in the area, the Mud Gap Trailhead (4,480 feet) lies on the border of Nantahala and Cherokee National Forests. A three-mile round-trip trail on an old roadbed leads to Tennessee's Whigg Meadows, known for its flame azaleas, blueberries, wildflowers and distant views. After the USFS purchased this site of Babcock Lumber Company's logging camp, farmers grazed cattle and sheep on the meadows until the 1970s. The Whigg family had built a cabin here in the late 1800s, and it served as a shelter for caretakers of livestock and men in hunting parties before it was later removed. On a night-training flight in April 1945, a World War II B-17 bomber out of Biloxi, Mississippi crashed into the side of Whigg Meadows, killing all ten of the servicemen onboard.

At Mud Gap on July 16, 2005, one hundred supporters of a volunteer organization celebrated twenty-five years of planning, hard work and achievement. The new Benton MacKaye Trail (BMT) was finished. From

Caretaker's cabin for the Hooper Bald Lodge. *Mike Ingram, Revonda McGuire Williams Collection.*

Hooper Bald Lodge. *Mike Ingram, Revonda McGuire Williams Collection.*

THE HOOPER BALD HUNTING LODGE

By Mike Ingram, Graham County resident

In about 1910, George Gordon Moore had a vision. Whiting Manufacturing Company, of which Moore was an agent, had acquired a large expanse of virgin timberland in the Snowbird area of Graham County that included the south side of Hooper Bald Mountain. Whiting intended to log the tract for lumber, but Moore decided that a shooting preserve with a European-style hunting lodge located near the top of the mountain would be a great place to entertain wealthy clients and influential friends. He arranged the lease of 1,600 acres from Whiting and procured funding for the project through English investors.

Using local labor, Moore next oversaw the construction of the ninety- by forty-foot log lodge. It would have ten bedrooms, a kitchen, a dining room, a large porch and a lobby and include the first indoor plumbing in the county. Moore also had a four-room caretaker's cabin built nearby and hired Cotton McGuire, a local resident, for the job.

Huge enclosures for the game animals surrounded the lodge and cabin, and by 1912, Moore had begun to bring them in. They included buffalo, bear, elk, Colorado mule deer, wild turkeys, English ring-neck pheasants and other fowl—even Russian wild boar imported from Europe. Moore may have assumed that such a variety of prey would be attractive to hunters, but it was hardly sporting, as the animals were penned in with high fences.

Springer Mountain, Georgia, through Tennessee and onto Davenport Gap, North Carolina, the three-hundred-mile BMT forms a five-hundred-mile loop with the A.T. The first graduate from Harvard's new forestry school in 1905, Benton MacKaye proposed the creation of the A.T. As a regional planner and co-founder of the Wilderness Society, MacKaye originally designed the A.T. to follow a more western route than hikers follow today. Now the Benton MacKaye Trail Association honors the conservationist's original vision.

From its juncture with the Cherohala Skyway at Mud Gap, the white diamond–blazed BMT heads north and passes a grave site marked, "Here

Moore himself made several trips to the preserve from 1912 to 1920, but the location was so remote that very few of his well-to-do clients chose to visit. After a long journey to reach the Snowbird Mountains, access to the lodge still required an arduous trek by mule-drawn wagons up steep slopes. Also, most of the imported, nonnative animals failed to thrive in the extreme environment, as winters on the bald could be severe. Others managed to escape the compound or were poached by locals. As a result, the venture was ultimately doomed.

Moore last visited the lodge in about 1922. Disenchanted with the failure of the project, he had moved on to become a lawyer and investment counselor in New York City, but Moore gave the caretaker, Cotton McGuire, permission to continue living in the cabin with an option to keep up the lease. McGuire and his family resided there for several years, farming, raising livestock and guiding the occasional party of local hunters. In 1938, the cabin was destroyed by fire, so the family briefly moved into the lodge, but by this time, the harsh winters had taken their toll. Burst water pipes, a stovepipe fire and a rotting roof had left it barely habitable. Eventually, McGuire, too, left the mountain.

The old lodge continued to deteriorate and was later bulldozed off the side of the mountain by a subsequent landowner. The preserve does have one lasting legacy, however. Many of the imported Russian wild boars escaped the enclosure and found the climate to be similar to their native environment. They successfully proliferated and still roam the Snowbird Mountains to this day.

lies an unknown man killed by the Kirkland Bushwhackers," a lawless, raiding party that roamed this area during the Civil War. After roughly following the ridgeline along the North Carolina–Tennessee border, the BMT crosses the Cherohala at the Unicoi Crest Overlook. Hikers continue north and cross Stratton Bald to a juncture with the Haoe Lead Trail. Here, Naked Ground and Hangover offer grand views. The BMT leads to Tapoco Lodge. Hikers cross US 129 to the James F. Burchfield Trailhead, named for a USFS official who fought to keep the Yellow Creek Mountain Trail section of the BMT open. After crossing the mountain, BMT joins the A.T. to enter the Great Smoky Mountains National Park at Fontana Lake.

In October 2015, near Mud Gap, "a gentle breeze fluttered American flags lining the Cherohala Skyway," wrote Kim Hainge, journalist for the *Graham Star*. One flag for each state guided distinguished guests and family members to a patriotic memorial service for nine U.S. Air Force airmen who had been aboard a C-141B. On August 31, 1982, during a low-level training flight, the Charleston-based cargo plane crashed into Johns Knob, less than one hundred feet below its summit, killing the crew. Phil Horne was scheduled to be on that flight. "I had missed a refresher class that was required annually and was told that I could not fly until I met those requirements," he told the reporter.

On Memorial Day in 2013, Horne had seen a sign to Tellico Plains while cruising the skyway with other motorcyclists. Painful memories of the fatal accident resurfaced. Until that day, he had never visited the crash site. After frustrating contacts with government officials, requesting a memorial marker for the soldiers who died there, Horne called retired USFS staffer Marshall McClung. McClung, a Cheoah District employee, had responded to the emergency scene thirty-three years earlier. McClung immediately garnered community support for the project. A few miles away from the crash site, private property owners donated the land for the monument. Williams Memorials provided the stone and arranged for etching. Cedar Cliff Baptist Church served meals. Honor guards, family members and local citizens attended the event.

"With tears in his eyes," reported Kim Hainge, Horne spoke at the event. "I am so grateful for a small patriotic town who would go all out to honor the families, friends and veterans in this great cause, unselfish in their donation of land, time, talent and resources to do this…a great example of an American small town of heroes answering the call to military heroes who gave their all to defend and protect this great land."

After the wife and son of one of the fallen soldiers received a folded flag, two buglers from the Monroe County Tennessee Honor guard concluded the service by playing "Echo Taps."

Farther west on the skyway, motorists cross Stratton Meadows Bridge near the Stratton Ridge Overlook with its wide parking area, restroom facilities and picnic tables. In the distance, 5,341-foot Stratton Bald caps the ridgeline. In the 1830s, John Stratton settled in this area and built a two-story log cabin in the Meadows. Due to harsh weather conditions and declining health, he moved to Tennessee in about 1858. Civil War drifters later burned the deluxe cabin. His son, John Wesley, for whom Johns Creek and Johns Knob are named, assumed ownership of the Meadows, grazing

MacKaye and I

By Dick Evans, member of the Benton MacKaye Trail Association Board of Directors for years and the 2011–12 president

When I retired in 2003 to the small town of Robbinsville, North Carolina, the Benton MacKaye Trail Association was proposing a three-hundred-mile wilderness-type trail in the area. I decided to join and get involved. After several work trips, I felt that I should learn a bit about the man for whom the trail was being named. I bought ***Benton MacKaye: Conservationist, Planner and Creator of the Appalachian Trail***, by Larry Anderson. In a quote cited by the author in the introduction, MacKaye mentioned Mulpus Brook—a very unusual name for a brook, I thought. My father, raised in the small hamlet of Shirley, Massachusetts, had mentioned a Mulpus Brook crossing the family farm in that area. As I read further, I realized that MacKaye grew up in the same town and kept returning to it after his various work projects around the country. I found it strange that my family had never mentioned him. I had an opportunity to visit an aunt and asked if she had ever heard of MacKaye.

"Ben! Oh, my!" she responded. She proceeded to talk for two hours about his local hikes (or "tramps," as he called them), his family and his life in Shirley.

One of my treasured keepsakes is a copy of MacKaye's book, ***Expedition Nine: A Return to a Region***, endorsed to my Aunt Ida Arnold in 1969. Although I have served on the board and as president of that fine group, the Benton MacKaye Trail Association, my greatest honor was to be chosen to chair the committee that planned and executed the "Trail Completion Ceremony." On July 16, 2005, along the Cherohala Skyway, we marked the culmination of MacKaye's dream to build a feeder trail from Springer Mountain, Georgia, to the Great Smoky Mountains National Park, connecting the Appalachian Trail at three points along its three-hundred-mile length.

large herds of hogs and cattle and leading hunting trips into the area. Another relative, Bob Stratton, for whom Bobs Creek and Bobs Knob are named, fell victim to the Kirkland Bushwhackers. Stratton and a friend had pursued their wayward cattle into Tennessee. Envious of Stratton's new breech-loading Spencer rifle, Kirkland's party ambushed the pair,

USFS Marshall McClung, who helped in the search-and-rescue operations for the plane crash on Johns Knob, identifies its location on a map. *Graham Star.*

killing Stratton with one shot and stealing his rifle. Today, the forest has reclaimed much of Stratton Meadows.

Beneath green-striped maples, alternate-leaf dogwoods, yellow birch and silverbell trees, a picnic table at Stratton Ridge sits beside Cherohala Skyway. Motorcyclists lean into the parking area, dismount and commune with others. Some read the informational exhibits near the center of the overlook. Others just compare last year's ride to this one. Parades of Miatas, MGs and Mini Coopers of all colors stream by, some topless. Between the loud blasts of motorcycles rocketing around Cherohala curves, a barred owl echoes from the distance—"Who cooks for you? Who cooks for you-all?" From Stratton Ridge, the skyway continues west to crest the Unicoi Mountains at the Tennessee state line. Increased rainfall and humidity cause a bluish haze to hang over the Unicois, or the "Unega," Cherokee for "misty" or "smoky."

SANTEETLAH LAKE

Northwest of Robbinsville, Santeetlah Lake stretches into long finger-like coves between tall mountain ridges, where its quiet waters offer peaceful solitude. NNF maintains 75 percent of its lakeshore. Between the lake's southeast border and Massey Branch Road, Cheoah District's ranger station sits near the confluence of Massey Branch and Cheoah River. A popular USFS boat ramp, a handicapped-accessible fishing pier and the local headquarters for the North Carolina Forest Service are nearby.

In 2016, across from the Massey Branch Ranger Station, the Partners of Joyce Kilmer–Slickrock Wilderness helped the USFS restore the historic site of Camp Santeetlah, CCC Camp F-24. Volunteers cleared an open meadow picnic ground and a memorial trail through a pine forest. They also rebuilt an observation platform. From 1937 to 1944, more than two thousand CCC recruits had camped here and planted trees, improved fish and wildlife habitats, built campgrounds and fought wildfires throughout the district. They upgraded Tatham Gap Road. They hiked through Yellowhammer Gap, carrying eight thousand brown trout to restock Slickrock Creek. When the USFS created Joyce Kilmer Memorial Forest, the CCC built its picnic facilities and nature trails. After recruits had barely completed a passable road in time for its dedication ceremony, overnight downpours saturated the plowed dirt. When guests arrived the next morning, their cars sank deeply into the mud. For nearly a mile, CCC men on tractors pulled stranded cars to the Kilmer Picnic area.

To build Haoe Bald Fire Tower, one hundred CCC men climbed more than three thousand feet over four miles. "Haoe Tower was a 14- by 16-foot toolhouse that had been moved a board at a time," wrote retired Ranger William Nothstein. "There was a 6- by 6-foot cupola like a church steeple on top, all glass, with a firefinder and telephone. The CCC boys carried every piece of lumber on their backs. You couldn't take a horse up, because one place was a stone step; you had to crawl to get over [it]." During construction, the men camped in tents on the high-elevation bald. They carried two gallons of water twice a day from Jenkins Meadow. From the Massey Branch Ranger Station to the battery-operated crank telephone at Haoe Tower, they cleared a phone line across Joyce Kilmer Forest and Jenkins Meadow. Heated with a wood stove and grounded for lightning, Haoe Bald Fire Tower served CCC and USFS fire wardens for years.

Leaving the CCC historic site, a left turn onto Massey Branch Road and then a right onto NC 143 leads visitors toward the Snowbird Community and a clockwise tour of Santeetlah Lake. Big Snowbird Creek and its tributaries tumble over Upper, Middle and Big Falls in the Snowbird Backcountry Area near the North Carolina–Tennessee line. Big Snowbird gathers waters from Sassafras Branch and flows east along Big Snowbird Trail. The trail follows part of an old narrow-gauge railroad bed from Big Junction Gap on the Cherohala Skyway to Junction at the end of FSR 75. At Junction, narrow-gauge rails met the standard-gauge line of the Buffalo and Snowbird Railroad, owned by Bemis Lumber Company and Champion Fibre Company (called the Champion Paper and Fibre Company after 1935). The five-mile-long Forest Service road retraces the old standard-gauge line. Winding along Big Snowbird Creek, outlined with spring wildflowers or fall colors, the graveled roadway offers a grandstand view of the stunning, dramatic play of light, rock and stream.

The Buffalo and Snowbird Railroad transported hardwood logs to the Bemis Sawmill near Robbinsville. Champion transferred its hemlock and pulpwood headed to its Canton mill to the Graham County Railroad. In 1925, the fifteen-mile-long Graham County Railroad sliced through a gap in the Snowbird Mountains, connecting Robbinsville with the Southern Railway's Murphy Branch at Topton. Engineered by Ed Collins, the original Shay locomotive, known as "the last steam freight line in the country," continued to haul loads into the 1970s and is now on display at the North Carolina Transportation Museum at Spencer. In the 1960s and '70s, the Bear Creek Junction, a scenic steam passenger train, pulled by one of the old Shay logging engines, also followed this rail line to Nantahala Gorge.

Haoe Lookout Tower and fire cabin, Cheoah Ranger District. *National Forests of North Carolina Historic Photographs, D.H. Ramsey Library, Special Collections, University of North Carolina–Asheville.*

In the early 1900s, Kanawha Hardwood Company logged the Snowbird watershed. It cut timber on Little Snowbird, Atoah and West Buffalo Creeks. When Kanawha sold its mill at West Buffalo, the company built a circular mill on Little Snowbird Creek and, reportedly, the first hydroelectric plant in Western North Carolina. According to Robert B. Barker, an Andrews historian and son of Kanawha's general manager, John Q. Barker, Kanawha provided electricity to this remote region, while "Asheville only had kerosene or gas lights." Kanawha also built the narrow-gauge Snowbird Valley Railway south to Andrews. Its president and its company doctor responded to calls by pedaling the line on four-wheeled railway cycles. Since many Snowbird Cherokees worked for Kanawha, William A. Lewin, president of Kanawha Lumber Company, provided hymnals printed in their own language. During World War I, France purchased the rails of the deserted Snowbird Valley Railway.

Big Snowbird Creek converges with the Little Snowbird in the valley where the Dentons managed Waldo Ranch and flows past the Snowbird Cherokee Indian School. Soon, it enters Santeetlah Lake near a USFS picnic area. Departing Snowbird Backcountry on FSR 75, Snowbird Road returns motorists to NC 143. Across NC 143 (Santeetlah Road), an exhibit marks the primary trailhead for the Santeetlah Lake Trail, created by the USFS, the GREAT organization and others. Many beginner-to-intermediate mountain bikers enjoy the challenges of the track, as do equestrians, hikers and runners. Volunteers help keep the trail system cleared of overgrowth, downfall and litter. Hike-the-bike sections cross hazardous portions of graveled roads. Some areas follow old FS logging roads, including one that runs through Long Hungry Campground with primitive lakeside camping sites. Much of the trail hugs the shoreline with fabulous mountain views.

Within a few miles, NC 143 crosses the prong of West Buffalo Creek, emptying into Santeetlah Lake. Three miles later, a driveway climbs to Snowbird Mountain Lodge. Almost surrounded by NNF, the lodge is listed in the National Register of Historic Places. In the 1930s, travel agents had promoted the health benefits and scenic beauty of the Western North Carolina mountains. A Chicago-based agent, Arthur Wolfe, arranged six-day tours throughout the region. After boarding the "Carolina Special" with Southern Railway, patrons left Chicago bound for Knoxville. Open-air tour buses drove from the Knoxville Depot to Gatlinburg, through the Smokies to Bryson City and Cherokee and on to Asheville. From the Grove Park Inn, visitors toured Mount Mitchell and the Biltmore Estate before traveling to the Tapoco Lodge on their way back to Knoxville. When Alcoa

"I am passionate about sharing trails with respect and kindness and commonsense," said Christine Vigue, Back Country Horsemen of North Carolina. *Christine Vigue.*

decided to end guest services at Tapoco, the Chicago travel agent and his brother decided to build their own mountain resort. In 1939, they came to Graham County.

Ed Ingram, a druggist in Robbinsville, suggested that they hire a local mountain boy as a guide—"natives are a little suspicious of outsiders." For fifty cents a day, the young man led the gentlemen around Santeetlah Lake. Near Joyce Kilmer, they found Wiley Underwood plowing his field.

"Know of any land for sale around here?" they asked.

"That mountain over there," he said, pointing to a 2,800-foot ridge above his farm. During the survey, the Wolfe brothers learned that eight of his wife's relatives owned the acreage. After clearing titles and finalizing the sale, the new owners chose a site for the lodge on a ridge facing the Snowbird Mountains. Downhill about one hundred yards, a spring provided fresh mountain water. The NNF district ranger offered help to build a phone line. When a power line from Nantahala Power and Light proved too expensive, generators provided power for lights and water. Twenty days of labor carved the driveway. Bemis Lumber Mill processed logs cut from the lodge site for the construction. After eight months, the Swiss chalet–designed mountain lodge opened for the spring season in 1941.

After the Wolfes sold the lodge in 1953, new owners added one of Graham County's first TV sets, and schoolchildren came to watch the county's first icemaker. Nine owners managed Snowbird Lodge over sixty years. In 1996, the year Cherohala Skyway opened, Robert Rankin purchased Snowbird Mountain Lodge.

"Men with vision created Snowbird Mountain Lodge," he wrote, "a vision of an inn with spectacular views, fabulous food, gracious hospitality and a peaceful setting. Caretakers have preserved it so guests will be able to enjoy her special treasures as we do every day."

With NNF as his backyard and as an advocate for the natural world, Rankin has been involved in national forest projects. He was a founding member of the Partners of Joyce Kilmer–Slickrock Wilderness. In November 1999, "Robert Rankin and the Snowbird Lodge saved the day," wrote retired District Ranger Lewis Kearney. Arsonists had ignited five wildfires around the Joyce Kilmer Wilderness area; 600 more acres were burning across the Tennessee line in Cherokee National Forest. In total, 5,600 acres were ablaze. At the Cheoah Work Center, Fire Incident Commander Kearney organized a command center; 360 firefighters and personnel arrived from eighteen states. Graham County was overwhelmed. Rankin offered Snowbird Lodge as accommodations at the local standard rate. Fire responders stayed in luxury rooms with a fireplace and hot tub. On Thanksgiving Eve, Rankin and his wife prepared a gourmet meal. An Arizona firefighter was amazed. "Is this the way you treat firefighters in the South?"

"Of course," was the nonchalant response from the NNF staff. "Y'all come back and help us anytime. This is the way we do it here."

On Santeetlah Road beyond Snowbird Mountain Lodge, NC 143 becomes Cherohala Skyway at Santeetlah Gap. North of the gap, SR 1127 leads to

MEDICINAL PLANTS OF THE SNOWBIRD CHEROKEE

By Ila Hatter, interpreter/naturalist for Great Smoky Mountains Institute; author of Edibles and Medicinals of the Southern Appalachia *and host of UNC-TV (PBS) Folkways program*

Blackberry (*Rubus allegheniensis*). *"Ka nv ga li."* Root tea for dysentery.

Boneset (*Eupatorium perfoliatum*). Tea for influenza, high fevers.

Ginseng (*Panax quinquefolius*). *"O da ga li"* or *"Yun wi Us di."* Added to formulas to make medicine more effective. Root chewed for stamina and to increase virility in men.

Horse nettle (*Solanum carolinense*). Roots strung into a necklace for a teething baby to relieve excessive saliva.

Jewelweed, Touch-me-not (*Impatiens capenis*). Crushed stems/leaves relieve poison ivy rash and bee stings.

Persimmon (*Diospyros virginiana*). *"Sa li."* Fruit pulp swabbed in a baby's mouth for "thrush."

Sassafras (*Sassafras albidum*). *"Ka na s dat si."* Root tea for blood builder, headaches, colds and a wash for poison ivy. Also, scented homemade soap.

Spicebush (*Lindera benzoin*). *"No dat si."* Twig tea for arthritis; parboil with small animals to remove "gamey taste" before baking. Dried berries used as substitute for allspice.

Sumac (*Rhus glabra*). *"Qua la gi."* Berries brewed into tart beverage or warmed for sore throat gargle. Wood makes smokeless fire. Also used for pipe stems.

Sweet birch (*Betula lenta*). *"A ti sv gi."* Twigs boiled for wintergreen flavored "aspirin."

Sweetgum (*Liquidamber styraciflua*). *"Tsi la lu."* Resin used to draw out splinters.

Toothwort (*Dentaria diphylla*). Crushed roots used to relieve toothache or headache. Leaves cooked in the spring.

Tulip poplar (*Liriodendron tulipifera*). *"Tsi yu."* Leaf buds crushed and mixed with lard or bear grease for a burn salve.

Yellowroot (*Xanthorhiza simplicissima*). *"Da la ne i."* Boiled root given for jaundice, urinary tract infection and stomach and mouth ulcers. Contains berberine, an antibiotic. Its bright color also easily dyes cloth or basket splits.

Witch hazel (*Hamamelis virginiana*). *"Tsu ki ni."* Known as "wheezel wood" for its inner bark tea for bronchial cough.

USFS Rattler Ford Group Campground with four campsites for groups up to twenty-five people. Horse Cove Campground, near the entrance to Kilmer Memorial Forest, provides tent and RV campsites. Santeetlah Creek entertains young waders before flowing into Santeetlah Lake three-fourths of a mile below the campground. Dozens of dispersed, primitive campsites dot the shores of the lake. Many are only available by boat. Managed by the USFS, primitive campsites beside the road overlook the lake. SR 1127 ends at the Maple Springs Overlook.

To continue around Santeetlah Lake, visitors travel SR 1134 toward the dam. A right turn onto SR 1147 leads to the USFS Cheoah Point Campground, with a boat ramp, a swimming beach, twenty-three RV/tent sites, picnic tables and hot showers. Although fly-fishermen enjoy the trout waters of Big and Little Snowbird Creeks, as well as Slickrock and Santeetlah Creeks, anglers at Santeetlah Lake cast their lines for large- and smallmouth bass, crappie, walleye, lake trout and bream.

In 1926–28, Alcoa built its 212-foot-high, 1,054-foot-wide Santeetlah Lake Dam to augment its hydroelectric needs. Alcoa, however, did not generate power at the dam. Through large steel and concrete pipes, lake water traveled underground, over roadways and down hillsides to a power plant at the old Rhymer's Ferry on the Little Tennessee River (now Cheoah Lake). A village with houses and a church, school and store accommodated powerhouse workers. When automated service from Cheoah Dam to Rhymer's became available, the village folded. In 2012, Brookfield Smoky Mountain Hydropower LLC bought Alcoa's Santeetlah Reservoir, Cheoah River and Cheoah Reservoir.

FONTANA LAKE

Alcoa built three lakes in the Cheoah Ranger District: Lakes Santeetlah, Cheoah and Calderwood. Perhaps it planned to build more reservoirs. Nonetheless, federal legislature mandated that Alcoa sell some of its land in the Little Tennessee River Basin to the Tennessee Valley Authority (TVA). TVA planned to impound the waters of Nantahala, Cullasaja, Tuckasegee, Oconaluftee and Little Tennessee Rivers by building the largest dam in the East. As part of TVA's water control system in the Southeast, Fontana Dam flooded homes, roads, railroads, villages and a copper mine. For recreational enthusiasts, it provided about 240 miles of shoreline and ten thousand acres of water surface.

After Congress created TVA in 1933, the agency built more than a dozen dams throughout the Southeast to provide flood control, river navigation and electricity. In 1941, when the United States entered World War II, the country needed TVA's help. TVA needed to furnish the power to manufacture war supplies at more than forty Alcoa plants, including the one at Maryville, Tennessee. At the nuclear weapons lab in Oak Ridge, Tennessee, the U.S. government entrusted TVA with a greater mission. With the building of the hydroelectric dam at Fontana and other sites, TVA would supply the kilowatt energy needed for the uranium enrichment to produce the world's first atomic bomb. Therefore, on the Little Tennessee River, Congress authorized the construction of the tallest hydroelectric, concrete dam east of the Mississippi.

TVA began buying more than one thousand tracts of land. The 480-foot Fontana Dam would flood Little Stecoah, Bushnell, Almond-Judson, Japan and other villages. Flooding more than eleven thousand acres of land, the approximately twenty-nine-mile-long and 130- to 430-foot-deep lake sent nearly six hundred Swain and Graham County families packing. Some stayed to build the dam.

The federal government closed the eleven-year-old copper mine on Eagle Creek that shipped its ore by Southern Railway to the smelters in Copper Town, Tennessee. Before the copper mine opened in 1931, the tent-town and village of Montvale Lumber Company had operated in this area. When the company left, Mrs. George Lidy Wood, the wife of Montvale's executive vice-president, left a perpetual signature, naming the area Fontana (Italian for "fountain") for the local waterfalls.

To ship its lumber to markets, Montvale's railway service had connected to the major hub at Bushnell fourteen miles away. Later, after purchasing thousands of acres of Fontana's forests, Whiting Manufacturing Company leased the railroad. Its train, named the "Big Junaluska" after the Cherokee leader, included a passenger coach with nearly thirty flatbed log cars. By the time TVA came to build the Fontana Dam, though, most of the logging companies were gone. In the late 1920s and early '30s, north of the Little Tennessee River, lumber companies, like W.M. Ritter Lumber Company at Proctor and Hazel Creek, left when President Calvin Coolidge created Great Smoky Mountains National Park.

By January 1942, six thousand TVA workers and their families had begun moving into Welch Cove about two miles away from the construction site. TVA's Fontana Village had nine hundred housing units, several churches, entertainment facilities, a store, a school, a hospital, a bank and a dining hall.

Military marches and broadcasts of patriotic music subconsciously urged dam workers to remain loyal to the cause, working three shifts a day, seven days a week. Above the doorway to the camp cafeteria, a sign read, "Work! or Fight!" reminding them of their national pride and duty in the war effort. With their dedicated allegiance and work ethics, men completed Fontana Dam in only three years.

During the construction of Fontana Dam, TVA diverted the Little Tennessee River through two concrete hillside tunnels that would later become the dam's spillway outlets. Today, when the lake has more water than the turbines and generators need to meet the electrical demand, TVA staff release water through these outlets. To minimize the potential erosive damage of water exiting these tunnels at nearly ninety-five miles per hour,

Fontana Lake. *Photo by author.*

The Fontana Dam Village housed families of the dam construction crew. *Illustration, Tim Worsham.*

concrete aprons built inside the tunnels deflect walls of water into 150-foot-long sprays that splash into the Little Tennessee 400 feet downstream. Water passing through the turbines producing the power for three massive generators rejoins the river below the dam.

Since steel was scarce during the war years, builders used reinforced concrete to construct the walls and framework of the powerhouse. Timber cut and cleared from the dam and lake sites became structural forms for the dam. Exploding 600,000 pounds of rock from the mountain's foundation with each blast of dynamite, men quarried stone about a mile below the dam site. Truckloads hauled thirteen tons of mountain fragments to one of two primary crushers. Secondary crushers pulverized stone into sand. Across the Little Tennessee River, conveyers transported aggregate materials to a concrete mixer plant.

In 1942, Fontana Village formed quickly. When workers finished the dam project in 1945, most of them left. Only TVA officials operating Fontana Dam remained. Discussions focused on turning the deserted village into a mountain resort. Nantahala National Forest bordered the lake and village to the south, and the Great Smoky Mountains National Park bordered their lake to the north. Like the lodges of Tapoco and Snowbird Mountain, Fontana Village Resort could offer respite and recreation in a forested mountain setting. Tourists began enjoying its wide range of activities, including fishing competitions, horseback riding tours and Square Dance Festivals.

Today, Fontana Village offers lodge and cabin accommodations, a lazy river and swimming pool, restaurants, concerts, guided hikes, recreational activities and holiday events. Buffets celebrate Thanksgiving, and fireworks blaze in July. The historic Jesse Gunter Cabin stands guard above Fontana's open greenway. By ox sled, Gunter had moved his family from Stecoah to Welch Cove in the 1880s. Fontana preserves the two-room structure with its full-sized loft and opens it for tours.

The marina at Fontana Lake rents boats, canoes and kayaks. Aboard the pontoon boat *Miss Hazel*, Captain Karl Sutter and others narrate guided history tours. Captain Sutter talks about Fontana's natural history, including the successful nesting of bald eagles on Eagle Creek, the medicinal plants of the Smokies and the bear sightings on Fontana's shoreline. In 2015, he and his passengers watched a bear cub swim across a narrow cove. He cruises by the island where *Nell* was filmed, the island that held the machine gun protecting Fontana Dam during the war, the site of the old copper mine and a rock cliff where locals take giant leaps. He describes submerged schools, churches, cemeteries, railroad depots and towns. The final history lesson culminates at Fontana Dam, idling past a diversion tunnel, the spillway and the massive dam itself. Onboard one of his tours a few years ago, Captain Karl met the son of a man who helped build the dam. As a crane operator, his father worked forty-eight-hour shifts. During his scheduled time, he ate, slept and worked up there until it was time to move the crane to another location. Between 1962 and 1972, when the United States feared a possible nuclear attack, Fontana Dam, according to Captain Karl, was one of four places in the country designated as a safe haven for government officials. Long tunnels and rooms were set up with foods, beds, blankets, desks, water purification and communication systems and other supplies.

Fontana Lake. *Vintage postcard.*

Fontana's hiking trails climb ridges, dip into coves and follow old rail lines of the Whiting Lumber Company. A path off the Whiting Rail Trail, called the "Stairway to Heaven," climbs the mountain to Lookout Rock, where it becomes part of the Benton MacKay Trail in NNF. The Whiting Trail, however, continues below the ridgeline and drapes along steep hillsides through an open forest, like a ribbon edging the folds of a curtain. Black-throated blue warblers slowly slur soft phrases, while chickadees loudly chide with scolding chatter. In the distance, a woodpecker drums a steady echo. Like whipped cream spread across scoops of ice cream on a banana split, cumulus clouds pile on top of the ole' Smokies on the southern horizon.

Like the Benton MacKaye Trail, other trails in NNF pass through the Fontana area, such as the A.T., which joins the BMT to cross Fontana Dam and enter the Smokies. Like Franklin and Hot Springs, North Carolina, Fontana received the designation of an official A.T. Community, protecting and promoting the famed trail. On March 26, 2015, the Appalachian Trail Conservancy, Mayor Houston of Fontana Dam, USFS supervisor Kristen Bail and others celebrated the event. Fontana joined the current partnership of more than thirty-six designated communities along the A.T. After receiving a $200,000 grant, the USFS began restructuring its Fontana Trail System in 2015.

At the Fontana Post Office, Lisa Whaley meets thousands of A.T. hikers each year. "Sometimes families call to ask where they can send supplies to hikers," she said. "Some hikers mail supplies to themselves and pick them up here. Right now, I have about seventy A.T.-hiker packages. I'll hold them for thirty days." At times, a hiker's Fontana-stamped postcard sends an update on their trail travel to friends back home.

In 2015, a German female hiker seemed distressed when her package had not arrived at Fontana. Many hikers have new boots mailed to them along the grueling journey, but Niki Rellon was expecting a new leg. "She was an amputee," said Lisa. In Germany, Niki had won the women's welterweight kickboxing championship. Now a resident of Colorado, she had biked from Alaska to Mexico City and backpacked the Pacific Coast Trail. However, fifteen months earlier, a forty-foot fall in a rappelling accident split her climbing helmet in half, broke her sternum and several ribs and shattered her left leg and hip. Sponsored by Adidas, Niki planned to be the first female amputee to thru-hike the A.T. Befriending her on Facebook, Lisa followed Niki's adventures. By the fall, she had accomplished her goal.

In 2016, Lisa and others on Facebook followed Swain County native Steve Claxton as he thru-hiked the A.T. to raise money for the Big Brothers/Big Sisters Program. His friend Case Hooper, a Robbinsville High School student, adopted the mission as his senior project. When Steve passed through Graham County at Stecoah Gap on the A.T., his friends shuttled him to Robbinsville High for a fundraiser event, hosted by Case in Steve's honor. A dinner, auction and talent show raised nearly $5,000 for Big Brothers/Big Sisters. On August 23, 2016, Claxton, who hiked under the name "Mustard Seed," summited Mount Katahdin in Maine, raising more than $40,000 for the youth organization that provides positive adult relationships for the next generation.

Staffed by volunteer TVA retirees, the visitors' center at Fontana Dam, with its videos, maps and exhibits, provides a wealth of information about the dam's operation and history. Diagrams explain how a hydropower plant works. Audio tapes play personal stories told by "dam kids" who grew up in the worker's town of Fontana. Displays discuss how the organization monitors the ecological health of its aquatic environments. To protect more than two hundred different types of fish that live in the Tennessee Valley system, biologists and water resource specialists annually monitor fish totals, adjust water height during spawning season and improve oxygen levels below the dam.

In the Little Tennessee and Hiwassee River systems, the rare sicklefin redhorse is a fish of special concern. The sicklefin, abundant in the past, was a reliable source of food for the Cherokees. Pollution from fertilizer and wastewater runoff, as well as oil and gas dumping, has reduced its numbers significantly. Collaborating with the U.S. Fish and Wildlife Service, NCWRC, Eastern Band of Cherokee Indians and others, TVA specialists are working to improve habitats and public awareness. To restock affected streams, some groups are collecting eggs and raising sicklefins in fish hatcheries. Activists promote the program "Shade Your Stream." By planting streamside trees and shrubs, residents can reduce runoff, erosion and water temperatures.

For fifteen years, Clint and Ina Horton have volunteered at Fontana's TVA Visitor Center. After thirty years with TVA and retiring as a powerhouse operator at Kentucky's Shawnee Fossil Plants, Clint converted his lifelong work experiences into a valuable resource for public education. His knowledge helps others learn about TVA's history and its many projects in the Appalachian region. As he leads visitors around the exhibits at Fontana Dam, his enthusiasm becomes contagious. When a naïve researcher of NNF comes armed with dozens of questions, he smiles and patiently says, "I'm enjoying this. I love answering questions. I could talk about the history of TVA all day." After discussing Senator George Norris, hydroelectric power, floodgates, turbines, deflectors and the Manhattan Project, Clint shared a light-hearted piece of Fontana Dam history—its only imperfection. During construction in 1944, some fun-loving worker pressed a nickel in its wet concrete, leaving a permanent nickel-headed imprint in a wall near the parking lot.

TSALI RECREATION AREA

Reportedly, more than three dozen versions of the tale of Tsali exist, including depositions, newspaper accounts, historical novels, letters, diaries and more. Military accounts differ from oral histories and Cherokee narratives. Written documents are not consistent. Embellished reports and personal sentiments emphasize some details more than other stories do.

Most accounts agree that near the end of the roundup for the forced march, William H. Thomas, the white brother of the Cherokees, joined Second Lieutenant A.J. Smith and his soldiers to capture twelve Cherokees, including the family of Tsali and his wife, Nancy, near the confluence of the Tuckasegee River and the Little Tennessee. Years earlier, Chief Yonaguska (Drowning Bear) had adopted young Will, making him a member of the tribe. As a grown man, Thomas fought to acquire legal land rights for his extended Cherokee family. Since Thomas had developed a good relationship with the Cherokees, it was thought, perhaps he could help negotiate a peaceful outcome for Tsali and his family.

An unfortunate skirmish occurred around November 1, leaving two soldiers dead and another injured. Tsali and a few family members escaped. With the help of Will Thomas, Colonel William S. Foster and his men searched for the escapees. Bands of Cherokees led by Euchella and Wachacha also helped, some historians say, because they believed that they could stay in the Appalachians if they assisted the search parties. A detachment located and executed Tsali's sons and, later, Tsali himself. When the soldiers under Colonel Foster left the mountains, they decreed that all Cherokees remaining there would be allowed to stay.

On Fontana Lake in the late 1980s, the USFS created the Tsali Recreation Area. In the year 2000, it added a campground with forty-two sites and hot showers. A boat ramp now provides access to the lake. With a permit, fishing is available. In season, hunting is allowed. On the wooded Tsali Peninsula, day-use parking fees permit use of the area's forty-mile, four-loop system of trails for bikers, horseback riders and hikers. For safety reasons, bike riders and equestrians alternate days of use. Constructed in the 1990s, the single-track, hard-packed trails cross creeks, climb ridges, cruise flatlands, traverse old logging roads and pass former homesites. Overlooks offer outstanding views of Fontana Lake and the Great Smokies. Due to episodes of southern pine beetle infestations at Tsali, the USFS has cut trees in affected areas, leaving brushy vegetative zones. Other areas are managed with prescribed burns, making the peninsula one of the best grouse habitats in the Cheoah District.

"Tsali Trails will get a facelift," reported the 2010 USFS issue of *Carolina Connections*. "Funded by economic recovery dollars, contracted crews will improve the Tsali Trail System, a mecca for mountain bikers who flock here from around the world." Trail crews repaired eroded gullies, leveled steep sections and improved water drainage problems. In addition, the International Mountain Biking Association trail building school and staff with Nantahala Outdoor Center arranged for thirty-five workers to upgrade the trail system. On April 10, 2016, after a year of preparation, volunteers with Nantahala Area Southern Off-Road Bicycle Association (SORBA), in collaboration with the USFS, worked more than eight hours installing eighteen new color trail map exhibits at each trail intersection. On each map, a location-identifier number helps USFS rescue teams pinpoint the site of reported emergencies.

On NC 28, east of Tsali Recreation Area near the juncture of Nantahala and Little Tennessee Rivers, the Almond Boat Ramp on Fontana Lake occupies a historic site. There, in October 1837, army officials established Fort Lindsay, the northernmost North Carolina stockade built for the Forced Removal. The army named the fort after Colonel Lindsay, the first commander in charge of the Removal, later replaced by Colonel Winfield Scott. Preventing Cherokee natives from seeking asylum in neighboring Quallatown and in a small refuge in Haywood County, soldiers at Fort Lindsay captured more than one hundred natives from Alarka, Yellow Town, Nantahala and surrounding communities. By June 1838, the military had transferred the Cherokee prisoners from Fort Lindsay to Fort Butler (now Murphy). By July 4, Fort Lindsay had closed.

Without delay, white settlers moved in. The fort sat at the river crossing of a major route, the Tennessee–North Carolina Turnpike. Although the townspeople first named their post office Fort Lindsay, officials changed the name to Almond in 1890 with the coming of the Southern Railway. The construction of Fontana Lake destroyed half the town. On land managed by NNF, the Almond Boat Ramp with a small marina and cabins covers the other half. An exhibit sponsored by the North Carolina Chapter of the Trail of Tears Association describes the history of the fort. In all the removal forts in North Carolina, only the grave marker of Private William Constant, stationed at Fort Lindsay in 1838, leaves an aboveground historical reminder of the forts created for the Trail of Tears in this state.

In an updated 2006 study of North Carolina's Trail of Tears, archaeologists studied the current landscapes and the historical documents of twenty-eight sites. In 1986, Congress created the Trail of Tears National Historic Trail. By 1993, the National Trail of Tears Association had formed chapters of the organization in nine states. Each chapter became responsible for its local trail studies and interpretive exhibits. In Western North Carolina, several groups—like the National Park Service, the North Carolina Chapter of the Trail of Tears, the Eastern Band of the Cherokee Nation, TVA, county commissioners, local citizens, the University of North Carolina Research Laboratories of Archaeology and the University of Tennessee Department of Anthropology—have joined this project. USFS archaeologists in North Carolina have assisted with the fieldwork. Some sites, like Fort Lindsay, are candidates for the National Register of Historic Places. Confined initially at Fort Lindsay and Fort Montgomery in the present-day Cheoah Ranger District, Cherokee families were eventually forced to leave their native homeland, climb their Snowbird Mountains and transfer to other detainment camps in the present-day Tusquitee District to await their journey west.

PART IV
TUSQUITEE RANGER DISTRICT

At the Asheville depot, Murphy-bound passengers boarded Murphy Branch rail line No. 17. Dubbed, in jest, the "Asheville Cannonball" for its ten- to fifteen-mile-per-hour speed, the train crawled west. After passing stations at Candler and Canton and crossing the Pigeon River, it headed past Clyde and Waynesville toward one of its most dramatic challenges, 3,315-foot Balsam Gap. Engineers used double-headers, a second steam locomotive behind the first, to help climb steep grades.

Since 1891, the Murphy Branch line provided service to secluded Cherokee and Clay Counties, but many had planned for it to operate much sooner. North Carolina had opened its Salisbury-to-Asheville rail service on the Western North Carolina Railroad in 1880 and proposed two extensions to the Tennessee border: one from Asheville to Paint Rock near Hot Springs and the other to Murphy and the copper mines at Ducktown, Tennessee. Crews completed the Paint Rock line on schedule. Unfortunately, the other line failed to meet its 1885 deadline to Murphy and never made it to Ducktown. Instead, the Marietta and North Georgia Railroad earned money from transporting loads of copper ore.

The Nantahala terrain and other factors hindered the Murphy Branch project. Organizers mismanaged funds. Engineers disagreed on the route. When one engineer tested the underground structure of Red Marble Mountain, a gooey white mud seeped through his borehole, forcing workers to reroute the track. Surveyors mapped an area; others mapped it again. Landslides destroyed the work of previous days or weeks. Convicts with hand

Southern Railway train no. 17 (Asheville to Murphy) at Willits, North Carolina, a railway stop between Addie and Balsam, on the Murphy Branch line, August 23, 1947. *Richard D. Sharpless Collection. Cherokee County Historical Museum.*

tools worked through winter storms, floods and rocky terrain. Some lost their lives. When a raft carrying a guard and Cowee Tunnel crewmembers capsized, nineteen shackled convicts drowned in the Tuckasegee River. Engineering feats, like the Cowee and Rhodo Tunnels and the forty-three-foot-high Hawknest Trestle, demanded arduous labor. After ten years, the crews completed the 120-mile Murphy Branch line, but its owner, Richmond and Danville Railroad, went bankrupt. Southern Railway assumed management and began transporting freight and passengers.

Beyond Balsam Gap and Sylva, westbound No. 17 passed Dillsboro and chugged through the Cowee Tunnel. With its tight squeezes, slight curves, rockslides and unexplained noises, some believed that the tunnel was haunted. The grave site of the convicts who drowned in the Tuckasegee River was nearby.

Past Whittier, the train's brakeman slowed No. 17's entry into the station at Bryson City and then rode the rails beside the Tuckasegee to its confluence with the Little Tennessee River at Bushnell. Bushnell was a major hub, connecting logging railroads from the Great Smokies to the Murphy Branch. It also provided an outside connection for hundreds of local families. During the construction of Fontana Dam, the Murphy Branch trains transported TVA inspectors and the supplies needed by the workers. Trains transported loaded boxcars to the dam site, where large hoses suctioned out cement. Without a wye at the dam, engineers backed the trains down to the building site. After crews completed the dam, the reservoir flooded the site of the former Bushnell station and its railroad tracks. A new Murphy Branch rail line was graded to continue west of Bryson City.

No. 17 continued toward Murphy, rolling past Almond to begin its scenic route through the forests of Nantahala Gorge. Near Wesser, the train traveled along the section of Nantahala River where TVA rerouted its course and dismantled two bridges across the river's horseshoe bend to reconstruct for the Fontana Dam project. In the 1970s, Nantahala Outdoor Center would open its whitewater business here. No. 17's passengers were unaware that Nantahala National Forest would eventually surround the railway route

Southern Railway's Murphy Branch rail line leaving the Cowee Tunnel and crossing the Tuckasegee River, where nineteen convicts drowned. *Southern Museum of Civil War & Locomotive History Archives, David W. Salter Collection.*

through the gorge and that portions of the rail bed itself would closely follow the boundaries of the Cheoah and Nantahala Ranger Districts. Past the little community of Nantahala, near Hewitt Station, early passengers would not have known that the mayor of Andrews, Percy Ferebee, would later donate these five thousand acres to the USFS. No. 17's passengers noticed the large talc mining operation before climbing the three-mile grade up Red Marble Mountain out of the gorge. At Topton, they crossed the Cherokee County line, entered the future Tusquitee Ranger District and descended the steep 4–5 percent grade down Red Marble Mountain.

At Topton in 1925, the Graham County Railroad connected to Southern Railway's Murphy Branch line. This spur line not only transported lumber from Snowbird, Santeetlah and other areas but also delivered food products and other goods to the citizens of Graham County. After its logging service ended, the Graham County Railroad served as an excursion line. "Children love the old-time steam engine at Bear Creek Junction Railroad at US 129 near US 19 at Topton," promoted a 1960s postcard. "The ride travels miles of scenic grandeur in Nantahala National Forest and overlooks Nantahala Gorge."

Murphy Branch No. 17 continued westbound. It puffed out of Topton and crossed a gap previously spanned by the Hawknest Trestle. During the heavy rail use of the logging period in the early 1900s, Southern Railway had filled in the gap and elevated the track bed. The Murphy Branch had provided a boost to the lumber industry when loggers connected their spur rail lines to its tracks and efficiently transported logs to mills or markets. After the Flood of 1916, the Murphy Branch provided a lifeline to the city of Asheville. The flood had destroyed all the railroad systems into Asheville except one. Only the Murphy Branch could deliver recovery supplies. Every hour of each day, a train departed Murphy for Asheville with shipments received from Atlanta via Marietta, Georgia. After its heavy use in rescue efforts, the Murphy Branch required extensive repairs.

No. 17 approached the Rhodo Tunnel, where the train's engine and mail car once derailed, killing the engineer and foreman. Today, the thunderous trip through the tunnel is uneventful. As the train entered the ten-mile-long, Andrews-to-Murphy Konnaheeta Valley, the travelers dozed with the train's rhythmic sway until it arrived at Andrews.

In the 1880s, Valleytown was renamed Andrews for Colonel Alexander Boyd Andrews, the Western North Carolina Railroad president whose leadership helped successfully complete the Murphy Branch. The busy Andrews Station and railroad brought industry, logging and prosperity to

Passengers await the arrival of Southern Railway's Murphy Branch at the Andrews Depot. *Cherokee County Historical Museum.*

the town. After logging companies set up businesses in Andrews, the town became a major hub for connecting railroads bringing wood to sawmills and tanneries.

North from Andrews, John C. Barker, the manager (then president) of Kanawha Lumber Company, graded a rough road through the Snowbird Mountains and into Graham County. He planned to pull wagons loaded with logs behind a steam tractor across his "Barker Road." Mud and mishap changed his mind. Instead, his Snowbird Valley Railway Company transported wood to an Andrews tannery. South of Andrews, the east–west ridge of the Valley River Mountains lies parallel to the Tusquitee Mountains. Circling an enclosed valley like two cupped hands held fingertip-to-fingertip, the two sister ranges now embrace the USFS Fires Creek Recreation Area. Southeast of the Andrews Station, spur lines from logging camps at Rainbow Springs and Buck Creek would also connect to Southern Railway's Murphy Branch. Portions of these routes are now NNF trails and roadways.

By the 1930s, the Great Depression and depleted forests had left tannery and lumber company workers unemployed. Some left. Others worked for the railroad. Born in a railroading family, one graduate of Andrews High School, Harold F. Hall, rode his railroad career to the top. His grandfather helped build the Murphy Branch. His father and three uncles were station agents. Starting as a telegrapher, Harold eventually became Southern Railway's president. His signature accomplishment was the merger of Norfolk and Western Railroad with Southern Railway in the 1980s to form Norfolk-Southern. When he died in Virginia at age sixty-four, his family buried him in Andrews. Along the old Murphy Branch line, "a special funeral train carried him home one last time," stated the engraved words on his grave site monument.

The engineer of No. 17 sounded the whistle, signaling Andrews Station passengers to get on board. The conductor called out the familiar phrase. Following the Valley River, the train continued westward over the wide fields of Konnaheeta Valley. Beneath the valley floor, the nearly one-hundred-mile Murphy Marble Belt widens from about a half mile in Swain County, North Carolina, to nearly three miles in Pickens County, Georgia. Since the mid-1880s, mining operations have quarried this corridor at two primary sites: one near Marble Hill and Tate, Georgia, and the other near Marble, North Carolina. In Georgia, the mines have mostly produced a pure white marble, used in many federal buildings, including the Lincoln Memorial. In North Carolina, the Columbia Marble Company quarried high-quality

In about 1910, twenty-three-year-old Weaver McLean of Swain County worked as a telegrapher at the Whittier Depot on Southern Railway's Murphy Branch. *Hunter Library, Western Carolina University, Cullowhee.*

The Columbia Marble Company processed marble at a community located along the Murphy Branch. *Cherokee County Historical Museum.*

white, pink and gray dolomite marble, shipping it to commercial markets on the Murphy Branch. The mine's specialty was Regal Blue, a blue and gray veined marble, used to build the current Cherokee County Courthouse in Murphy.

Beyond Marble, westbound No. 17 rolled into Tomotla, near the site of a former Cherokee village on the Valley River. Today, significant Cherokee populations in the county live near Tomotla and Grape Creek. Delaying No. 17's journey for about ten minutes, steam engines stopped at the last water tank on the line to refill the tender. Firemen swung out a long spout on an elevated water tank, allowing gravity to fill the tender with enough water to last until the return trip from Murphy the next day.

On this trip, no rocks, hogs or cows delayed No. 17's service from Asheville to Murphy, arriving on time at the Southern Railway Depot. Nearby, passengers could board trains departing the Louisville and Nashville Railroad Depot for other destinations. Murphy, named for Archibald D. Murphey (with an *e*), a state senator who fought for public education, lies at the confluence of the Valley and Hiwassee Rivers. In the 1940s, TVA dammed the Hiwassee River. Its floodwaters not only covered former homesites and villages but also some of the Murphy Branch trackage north of the Valley River at the station.

Near the old railroad station, the historic site of Fort Butler rests on a hill above present-day Murphy. Fort Butler was a military headquarters for the removal of the Cherokee people. Today, surrounded by a residential area, massive tree trunks spread wide crowns to shade the grassy knoll and commemorative marker. The military encampment, which presumably covered several acres, had open-air stockades, a hospital, a blacksmith's shop, a quartermaster's office, a surveyor's office and other structures.

The three thousand Cherokees evicted from North Carolina passed through Fort Butler, "one of the most historically significant resources associated

The Murphy Branch rail service terminated at a depot near the Hiwassee River and the Louisville & Nashville (L&N) Depot. *Cherokee County Historical Museum.*

with the Cherokee Removal in North Carolina," stated a researcher with the University of North Carolina Research Laboratory of Archaeology. The army gathered Cherokee families into four detainment forts and two camps located throughout the Nantahala region. Soldiers then moved each group to Fort Butler to prepare for the final march west to the Indian Territory. On June 18, 1838, the first of three groups of the Cherokee people departed Fort Butler. In Murphy today, the Cherokee County Historical Museum is the local interpretive center for the Trail of Tears National Historic Trail. The museum preserves more than two thousand Cherokee artifacts, and its exhibits detail local Cherokee history.

In 1839, North Carolina created Cherokee County from lands in Macon County. Businesses grew quickly. Farmers plowed Konnaheeta Valley. The Murphy Branch rail line boosted the economy, but within decades, hard-surfaced roads, delivery trucks and privately owned vehicles had reduced rail travel. In the late 1940s, the Norfolk-Southern terminated passenger travel to Murphy, and by the late 1980s, it had stopped its freight service west of Dillsboro as well. Today, more than 200,000 passengers a year all come aboard the Great Smoky Mountains Railroad to follow the historic route of Southern Railway's Murphy Branch from Bryson City to Nantahala Gorge.

CHEROKEE COUNTY

The Tusquitee Ranger District is located in Clay and Cherokee Counties. Its ranger station lies near the Hiwassee River in Murphy. A tour of the district in Cherokee County begins with a clockwise drive of the Hiwassee Lake area. Leave Murphy on US 64 and cross the Nottely River, which joins the Valley and Hiwassee Rivers to form Hiwassee Lake.

"Tain't fittin'," said Old Ben to the Tennessee Valley Authority. In *The People's Giant: The Story of the TVA*, Harry Edward Neal recorded Ben's conversation. The agency wanted to take Old Ben's home. It planned to remove many residents along the Tennessee River and its tributaries from Knoxville, Tennessee, to Paducah, Kentucky, to build several large dams. Some homeowners accepted TVA's monetary offer willingly and moved away with or without TVA's help. Others balked, forcing TVA to evict them by eminent domain. A few committed suicide. Those unhappy with TVA's price for their property took the matter to court. For Old Ben, though, the matter just "tain't fittin'."

"My pappy's pappy built this house," he told them. "My pappy was born here and so was I." He pointed to flames glowing in the fireplace. "That there fire has been burnin' night and day since I came into this world," he told the TVA, "and them embers ain't about to die out whilst I'm alive!" Officials found a higher place on a hill for Old Ben and moved him, his cabin and his hot coals to the new site.

Along the Tennessee River system, TVA began moving residents soon after President Franklin D. Roosevelt signed the bill in May 1933 creating the Tennessee Valley Authority. The new agency planned a unified dam development project that would reduce flooding, promote regional economic development, generate hydroelectric power, control malaria by eliminating mosquitos breeding in stagnant water and provide a navigational route along the Tennessee River's 650-mile course. Like a long, narrow stair-stepped terrace, the construction of multiple dams along the river's main stem and several major tributaries created a sequential set of reservoirs dropping in elevation along its route.

On the Clinch River, a tributary of the Tennessee, TVA built its first dam in the unified project in 1933. Norris Dam, named for Senator George Norris, who fought for the creation of the TVA, began operating by 1936. In the fall of 1936, TVA transferred the Norris construction site bridge to its construction site for the Hiwassee Dam, its sixth dam within the Tennessee River Basin.

Surveyors studied the area to determine the best site to construct the Hiwassee Dam. They researched property boundaries and land titles. They drew contour maps of the area's steep terrain. The U.S. Army Corps of Engineers captured the region's topography in more than 350 aerial photographs. Geologists studied the area's rock foundations. After reviewing the compiled data, TVA chose to build its dam on the Hiwassee at Fowler's Bend, named for a family whose ancestors had lived there since the 1850s.

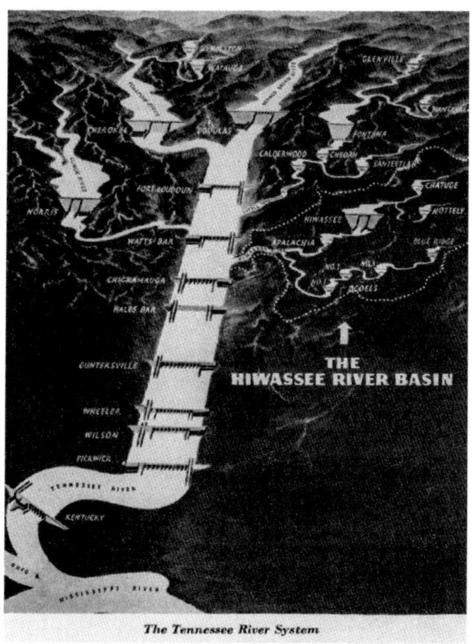

The Hiwassee River Basin. *From "The Apalachia, Ocoee No. 3, Nottely and Chatuge Projects,"* Hiwassee Valley Projects Technical Reports, *vol. 2, Government Printing Office, 1948.*

Over two and a half years, TVA purchased about twenty-three thousand acres. It acquired almost half of the acreage from a power company. The agency obtained the rest from about 260 families, mostly farmers. Farming and timber harvesting were the primary means of income in Cherokee County at the time. In its 1935–36 sociological survey of Cherokee County for the Hiwassee Project, TVA stated, "Until recently the primary source

An ox-driven cart pulls a load of railroad ties up Winding Stair Road, August 13, 1946. *National Forests of North Carolina Historic Photographs, D.H. Ramsey Library, Special Collections, University of North Carolina–Asheville.*

of income in the county was from the sale of sawtimber….[T]he present stand is mostly second growth not yet suitable for market except as acid or pulpwood, cross ties and telephone poles. Eleven percent of the county area was reserved as national forests in 1935."

More than two-thirds of the residents accepted TVA's offer and moved to other areas in the region. Altogether, those who refused to move lost about eight hundred acres to the power of eminent domain. Many local farmers, sawyers and others welcomed the construction jobs and steady income offered by TVA. The agency also provided family housing, a school, a cafeteria and dormitories for unmarried men near the construction site.

TVA built several roads and bridges and a 1.7-mile spur line to the Louisville and Knoxville Railroad to transport supplies. It cut thousands of forested acres and relocated churches, stores and more than four hundred graves. Steam shovels and bulldozers cleared the dam site for crews to build three cofferdams that diverted the Hiwassee River while men constructed the dam's foundation. By February 1940, TVA had completed the 307-foot-tall concrete Hiwassee Dam, which, at the time of construction, was the highest overspill dam in the world. A highway now stretches across its 1,376-foot width, providing a grand view of the lake. The powerhouse sits at the base of the spillway. After TVA completed the dam, it relinquished more than seventeen thousand of its surrounding acres to Nantahala National Forest.

West of Murphy on US 64, motorists turn right onto NC 294 and take another right on NC 1303, which eventually splits into FSR 85 and FSR 307. FSR 85 heads toward the Panther Top Fire Tower and the Panther Top Shooting Range. In 2005, the shooting range opened after three years of preparation and construction to offer a safe and legal environment for firearm practice. Volunteers built targets for both pistol and rifle ranges with right- and left-handed shooting benches.

In 1940, the year TVA completed the Hiwassee Dam, the CCC built a fire tower on 2,239-foot Panther Top Mountain. During construction, the men at Camp F-29 near Murphy rode army trucks to the mountain each day. Young Eddie Graham stood by his home on Panther Top Road, handing out bags of apples as they passed. In exchange, the men tossed him company-issued rations.

"I was seven years old when the CCC built this tower," said USFS volunteer Eddie Graham. During the last weekend in October 2016, he and his cousin Kenneth Graham offered guided tours of the thirty-foot-tall structure, listed in 1997 on the National Historic Lookout Register. Like fire wardens decades ago, visitors watched that day as four wildfires burned in the Nantahala region.

"I spent some Friday nights on Panther Top with Grady Waldroop, its first fire warden," said Eddie. During his USFS career, Waldroop also worked at the Wayah Bald and Albert Mountain Fire Towers. Waldroop slept on the south-wall cot, while Eddie took the west-wall bunk. Above Eddie's cot, a hinged wooden door held a large map. An Osborne firefinder sat in the center of the room. When Waldroop spotted smoke, he turned its sights toward the fire to take a compass bearing. He phoned his findings to the local dispatcher, who alerted wardens at the Big Stamp Fire Tower in the Valley River Mountains, the Albert Mountain Fire Tower in the Nantahalas and the Brasstown Bald Tower in Georgia. Each tower operator took compass readings of the fire from their positions. Grady and young Eddie then marked each warden's line of sight with strings on the fold-down map. The lines crossed or triangulated the fire's exact location. Today, Panther Top's original Osborne firefinder stands like a monument in tribute to its past guardians of the forest and is one of the few firefinders preserved in the tower that it served.

Outside, near the tower, a cistern collected water for bathing and cleaning. About a quarter of a mile below the summit, Waldroop and Eddie collected drinking water from a cold mountain spring. Today, partially hidden by brushy overgrowth, an old wooden latrine rots at the edge of the summit.

In about 2002, federal dollars helped restore the Panther Top Fire Tower. Inside, contractors repainted the walls and resurfaced the floor. They installed custom-made windows, constructed with materials that matched those used by the CCC. Outside, they restored the roof and fashioned new gutters to resemble the original ones. On a concrete pedestal below the tower, a plaque bears the names of the eleven men who served on Panther Top.

Beech Creek Genetic Resource Management Area

At the end of NC 1303, a left turn onto FSR 307 leads to the USFS Beech Creek Genetic Resource Management Area (GRMA), known simply as the "seed orchard." In the 1940s, scientists experimented with genetically improving the frequency and abundance of a plant's seed production. They also learned to breed, or clone, trees to produce desired results. Geneticists and foresters cut three-inch-long limbs, or scions, from mature parent trees with exceptional qualities, like a straight trunk, few lower limbs, greater disease resistance and more timber volume. They grafted the scion onto a healthy, two-year-old rootstock to supply water and nutrients to the new tree, which produced offspring with the superior characteristics of its parent tree.

"Grafted trees produce seeds earlier than native trees," said USFS Cheoah/Tusquitee fire management officer Chad Cook during a guided tour of the Beech Creek seed orchard. Within two to four years, the pollinated flowers on the new tree can produce seeds. In a natural forest, some trees may take twenty to twenty-five years to reach seed-bearing age. At age fifteen, Cook began working with the North Carolina State Forest Service as a firefighter, raking leaves and washing vehicles. Harold Norton, the last Panther Top Fire Tower operator, was a co-worker. Later, the USFS Tusquitee District hired Cook as a forest technician, helping, at times, in the seed orchard. After twenty USFS years, Cook now works in fire management.

"Early tree breeders showed that traits important to timber and pulpwood were inherited and could be improved to increase forest productivity and wood quality," stated the 2013 USDA report *The Current*

Seed Orchard Techniques and Innovations. Seed orchard policies also address USFS ecosystem management principles, such as producing mast (like acorns) as a wildlife food source and preserving endangered and threatened tree species. To supply seeds for the reforestation of federal lands, the USFS established seed orchards across the country. In the Northwest, managed orchards produced seeds for the reforestation of Douglas fir, western hemlock and other species. In Nantahala National Forest, the USFS established the Beech Creek GRMA in the 1960s, producing pine and hardwood seeds for the reforestation of the eight national forests of Kentucky, Virginia, North and South Carolina, Georgia, Tennessee and parts of West Virginia. In 1998, the Beech Creek staff also assumed the management of the Chilhowee GRMA in southeastern Tennessee. After Hurricane Hugo devastated the Francis Marion Seed Orchard in South Carolina in 1989, destroying the only seed source for longleaf pine population, the Beech Creek orchard began to duplicate, distribute and preserve genetically improved seed sources at multiple orchards.

In the forty-three southeastern ranger districts, forestry staff identify potential parent trees. Professional foresters, or tree markers, then evaluate the trees using established criteria for premium qualities and mark the approved specimens with a painted band and number. Staff then cut scions from the approved parent tree, graft them onto rootstocks and place them in a designated plot in the seed orchard, called a geographic seed source. Each seed source contains seven to sixty family members, or clones, from that parent tree, each bearing a number that represents its parent's location in the forest and the new young tree's individual identity.

At the 250-acre Beech Creek GRMA, each tree species from a specific location grows in a separate plot. For instance, the Kentucky shortleaf pine grows in a seed source plot separate from the North Carolina shortleaf pine. The seeds produced in each plot can return to their identified home region. Two hundred Beech Creek acres support source plots for four conifers: the Virginia, pitch, shortleaf and white pines. The remaining grafted clone-banks hold six species of hardwood trees: red, black, white and chestnut oaks; black cherry; and yellow poplar. Established in the mid-1990s, the northern red oak and white oak each occupy a seedling seed-source block. Two genetic conservation areas help researchers study the American chestnut and butternut trees. A plot for the butternut, or white walnut, a tree listed as "a species of concern," preserves the genetic material of this uncommon species. In a plot of flowering dogwoods, scientists study the effects and progression of the Anthracnose disease.

Once the geographic source plots are established, they become "first generation seed orchards." Then orchard managers and others initiate a tree-breeding program on each family plot to prove that the superior traits of the parent tree passed to its progeny.

"Genetic gains are determined in the progeny by comparing the genetic characteristics of the original parent clones," explained USFS Robin Taylor, current Beech Creek orchard manager. "Clones are hand-pollinated from a bucket truck using several pollination techniques. Paintbrushes and squeeze bottles are the most desirable ways to introduce pollen to receptive flowers isolated in a pollination bag. The bag stays in place until the flower closes and no chance of pollination from outside sources can occur. [On pine trees,] the resulting pollinated flower becomes a conelet and develops over a year into a mature cone."

Staff attach identification tags to the cones, collect the cones when they ripen and extract the seeds that grow into seedlings. Progeny tests begin on the new one-year-old seedlings and continue every five years until age thirty. The accumulated data determines the best family members in each clone-bank, essentially the best offspring of the superior-rated adults. Staff then graft these chosen first-generation champions into second-generation seed orchards, and the progeny-testing regimen begins again.

Harvesting the seeds from the Virginia pine "was a painful process," said former USFS seed orchard manager Glen Beaver. "It eats your hands up."

Clay Logan, a retired USFS seed orchard technician and the owner of Clay's Corner Store near Brasstown, known for its New Year's Eve "Possum Drop," recalled the same painful, labor-intensive process. The cone, covered with sharp spines, required workers to cut off each one individually from the limb and extract its seeds one by one with hand clippers. A portable cement mixer simplified the process. Cones tumbled inside the screen-covered, rotating barrel, dislodging the seeds.

Crews harvest white pine seeds when they ripen in late August to early September. The old method required hand-collecting 2,500 bushels of pinecones, each bushel containing about a pound of seeds. "It was awful work," said one worker, shaking his head, and "nasty, sweaty work."

Today, a tractor driver unrolls a fifteen-foot-wide bale of netting material between the mile-long rows of trees. Workers staple the fabric seams together, pulling the net close to the base of each trunk. A tree shaker machine vibrates the trunk, and a massive amount of seeds with attached seed wings fall onto the nets. Then the net retrieval machine and seed collection system begins its work. Bits of twigs and cones drop onto a conveyor belt to a shaker tray. Vibrations separate the seeds, which drop into collection bins.

During the drive through the Beech Creek Orchard, four deer watched as the USFS truck passed. Deer, birds and other wildlife often compete for the valuable seeds. Acorns are favorites for deer and bears. Beneath rows of oaks, streamers tied to tin pans twisted in the wind above collection tubs and groundcover nets to discourage deer from foraging on prize acorns. Flocks of birds perched in trees, waiting for seeds to mature. Once, an innovative employee attached a propane tank and a bottle to a timer. He devised the automatic system to blast like a cannon to frighten the birds. Startled at first, the birds soon grew wise and ignored the silly prank.

Orchard staff tend the tree plantation like a garden by weeding, fertilizing and spraying for diseases and insects. Since grafted trees are not growing on their own roots, they are often more vulnerable to destructive insect pests. During the 1970s and '80s, about ten members of the Youth Conservation Corps and the Older Americans Program worked twenty-four hours a week to help with seed orchard duties. Cook's seventy-year-old grandfather, H.L. Martin, who became the foreman for the Older Americans orchard crew, hand-pulled weeds, now controlled with chemical sprays.

The collected seeds are cleaned and dried to specified moisture content and then stored at the Ashe Extractory in Brooklyn, Mississippi. After USFS staff in the Southeast place an order for requested seeds, the seeds are grown in contract nurseries into seedlings. The North Carolina State Forest Service will often purchase excess seeds in stock. Some USFS dogwood seeds are grown in state orchards into seedlings and transplanted along state highways. The University of Georgia, the University of Tennessee and other institutions conduct research studies at Beech Creek GRMA.

HIWASSEE LAKE

Southwest of Hiwassee Lake on NC 294, a sign directs visitors to the USFS Cherokee Lake Recreation Area. In the 1940s, the CCC dammed Persimmon Creek, a tributary of Hiwassee River, to create the small lake. From April to October, the day-use area is open for fishing or to enjoy a picnic. An easy hiking trail leads to the historic dam.

Farther west on NC 294, a turn onto NC 1314 leads to the Hiwassee Dam. En route, a USFS sign identifies a launch site at Micken Branch where boaters and fishermen access the six-thousand-acre lake. Anglers fish for bluegill, large- and smallmouth bass, white bass, walleye pike and others. The narrow, twenty-two-mile-long reservoir curves around Ramsey Bend and John Green Bend before reaching Fowler Bend at the dam. Hiwassee Lake, wrote Bill Sharpe, is "so confined in a maze of mountains and ridges that it penetrates a good portion of western Cherokee county, writhing and twisting across the map like a dying dragon."

Near the dam, the highway passes the Bear Paw Resort, the former TVA village that housed about one thousand dam workers from 1937 to 1940. Beyond the guardhouse and security gate, Bear Paw's Village Drive climbs a hill past restored TVA cottages and tenant homes, often rented to guests by the owners. The agency moved some of the temporary houses to its new project at Fontana Dam. Also available for rent, the former TVA superintendent's home inside the complex still overlooks Hiwassee Lake from its mountaintop setting.

During World War II, the U.S. government promoted construction of dams to meet the increased hydroelectric needs of the war effort. In the Hiwassee River Basin, TVA built the Hiwassee, Apalachia and Chatuge Dams in North Carolina. Beside the Hiwassee reservoir, U.S. Navy personnel began occupying TVA's Hiwassee Dam Village in 1942.

For eighteen months after the United States entered World War II, frustrated servicemen used defective torpedoes that failed to reach their targets, exploded prematurely or never exploded at all. In a few cases, the torpedo's direction control locked, curving the missile back toward its own submarine. Theodore Roscoe, who penned the official U.S. naval and submarine history, stated, "The only reliable feature of the torpedo was its unreliability."

Setting up torpedo testing facilities across the nation was essential. One secret military location was Hiwassee Lake, with depths up to 250 feet. The military stayed in the former TVA housing units, including the former superintendent's cabin. Men also slept on the gym floor in the school. Ron Taylor, author of *A Bear Paw Ramble*, interviewed a former Boy Scout whose troop camped at the village on the weekends during the military's occupation. TVA's superintendent of construction was the scoutmaster; a member of the navy research team was his assistant. The navy allowed the Boy Scouts access to the military boat launch on the lake, shortening the long hike to a remote campsite.

With army guards stationed on the dam and the Coast Guard patrolling the waters, the Naval Ordinance Laboratory Experimental Facilities at

Hiwassee Dam launched test missiles from concrete pads on the hill above TVA's old water tank. Underwater cameras captured the action. Steel nets strung across the cove retrieved the cigar-shaped explosives. According to TVA, the national defense projects at Hiwassee continued for nine years, ending in the 1950s. Concrete reminders of Hiwassee's participation in its U.S. naval service remain at the Bear Paw Resort today.

Across from the entrance to Bear Paw, a paved road leads to the powerhouse at the base of Hiwassee Dam. Picnic tables are available. At the base of the dam, the Hiwassee River becomes Apalachia Lake, TVA's next reservoir in its stair-step, unified Tennessee Valley system. From Chatuge Lake at Hayesville to its junction with the Valley River at Murphy, the Hiwassee River drops 360 feet. From Murphy through the Apalachia Reservoir to the Tennessee state line, the river drops another 314 feet. About nine miles downstream from the Hiwassee Dam is the 150-foot-high Apalachia Dam located near the state line. An eight-mile-long penstock funnels the water farther downstream to the powerhouse. Although the remote, one-thousand-acre Apalachia Lake, bordered by NNF, offers little public access, kayakers and fishermen enjoy its solitude.

"My father worked for TVA when they were building Apalachia Dam," said Wanda Newman during an interview at the Fontana Lake marina. "He drove a Woody Ford wagon each day, transporting workers from Isabella, Tennessee." When TVA completed the Apalachia project, Wanda's family moved to Fontana, where she became one of the "dam kids," living with her parents at Fontana Village during construction. Between 1942 and 1946, more than six hundred students attended Fontana Village School, now the recreation hall. For decades, the dam kids have returned annually for a Fontana Village reunion.

"In 2015, there were only about seventy-five of us," she said. "We're getting older. But we still try to get here, talk about the good old days and catch up on the new. We laugh, we hug and, then, we cry when we have to say 'good-bye.'"

NC 1314 continues across Hiwassee Dam, past picnic tables and an observation platform, to the community of Violet. Right turns at two intersections circle back east around Hiwassee Lake along Joe Brown Highway (NC 1326). This route, formerly known as the Unicoi Trail (an ancient Indian footpath) and later widened as the Unicoi Turnpike (a major east–west wagon route), roughly follows a section of the Trail of Tears and has been designated a National Millennium Trail. North of here, the USFS has closed the Upper Tellico Off-Road Vehicle Area.

Apalachia Dam. *TVA.*

The Joe Brown Highway loses sight of Hiwassee Lake, crosses in and out of NNF and arrives at the USFS Hanging Dog Campground. There, a boat launch and fishing pier provide lake access. A low-water boat ramp for winter use is located nearby at the end of the peninsula at Ramsey Bend. Two camping loops offer thirty-four sites from May to September. A covered pavilion provides space for picnics. Before excavating the site for a latrine, the USFS conducted an archaeological assessment. Generations ago, the Cherokees occupied the Hanging Dog region on the banks of the Hiwassee River. Some hid in this area during the Forced Removal.

"We gathered hundreds of pottery chips at Hanging Dog before we could dig there," said Glen Beaver, a retired USFS employee who helped build the campground.

In the 1800s, miners discovered a large deposit of hematite (iron ore) near Hanging Dog. In exchange for a state land grant for three thousand acres, Harmon Lovingood agreed to establish a bloomery forge there. Acres of trees provided the charcoal for smelting iron. On May 6, 1865, a Civil War skirmish took place at Hanging Dog. Unionists had burned the Murphy Courthouse, possibly to destroy records of their misdeeds. Confederate soldiers, camping at Valleytown (Andrews), responded. The Blue and Gray met at Hanging Dog in one of the last Civil War battles.

Now, 150 years later, the area beckons tourists. "Pedal your way through the mountains along the water's edge," promoted a Tusquitee District brochure for the Ramsey Bluff Mountain Bike Trail system. Located near the Hanging Dog Campground, the course offers "under-utilized adventures in mountain biking, hiking and cross-country running." Eight miles of blue-blazed, easy-to-moderate trails loop together. Overhanging branches and trailbed rocks create challenges. About five miles beyond the Hanging Dog Campground, Joe Brown Highway joins Tennessee Street to return to Murphy.

CLAY COUNTY

A journey east of Murphy on US 64 follows another portion of the Unicoi Turnpike. Before the state built the turnpike in 1816, Cherokees traded along this route between their villages in the Piedmont of South Carolina and the mountains of North Carolina. Beside the turnpike in 1837, near present-day Hayesville, the U.S. Army conveniently built Fort Hembree. The fort, named for Joel Hembree, captain of the Tennessee Volunteer Militia assigned to apprehend local Cherokees, held Cherokee families awaiting the march west. The fort later became the area's center of business until Confederate soldiers occupied it. After the Civil War, the officers' quarters became a private residence. Removed in 1939, its bricks now form the foundation of Hayesville First United Methodist Church.

Clay County, created in the 1800s, is North Carolina's second-smallest county. (Chowan County is smaller.) Early settlers farmed the fertile valleys of Tusquitee, Shooting Creek, Hiwassee and Brasstown. Brasstown became the home of the historic John C. Campbell Folk School in 1925, preserving Appalachian craftsmanship and culture. After the logging and tannery industries moved away, more than half of Clay County became part of Nantahala National Forest.

Chatuge Lake and Jackrabbit Mountain Recreation Area

Eighteen miles east of Murphy on US 64 is Hayesville, a city, reportedly, closer to four other state capitals than its own state capital at Raleigh. A right turn onto NC 175 crosses Chatuge Lake and intersects with Jackrabbit Road (NC 1155). Brown signs direct visitors to the Jackrabbit Mountain Recreation Area. On a beautifully forested seven-hundred-acre peninsula, the USFS manages a campground and trail system on the banks of Chatuge Lake. In 1941–42, TVA built its 150-foot-tall dam across the Hiwassee River near the mouth of Shooting Creek, forming its seven-thousand-acre reservoir from southern Clay County into northern Georgia. Like other TVA projects, the agency moved churches and cemeteries to higher ground. The waters flooded house sites and a section of the former Unicoi Turnpike. In the 1940s, Chatuge Dam was the highest earthen dam in the world.

"Chatuge Lake is touted as the crown jewel of TVA's system of lakes," boasts USFS literature. As "one of the most complex, developed recreation areas in North Carolina," the Jackrabbit area lies near a bend in the thirteen-mile-long reservoir. Its mountain scenery is "similar to a Swiss Alpine lake setting." Each of the three campground loops offers comfort stations and hot showers. Open from May to September, the area also offers an amphitheater, a swimming beach, a picnic ground and a boat launch. During the lake drawn-down period from September to April, rockhounds comb the wide shoreline for Clay County minerals. In 2011, the USFS upgraded campsites and connected the local municipal water and sewer system to the campground.

The year 2011 also celebrated the grand opening of the Jackrabbit Mountain Bike and Hiking Trail System near the campground. After ten years of preparation and construction on a jagged, sister peninsula on Chatuge Lake, nearly fifteen miles of single-track trails now traverse forested landscapes and offer lakeside views. A three-mile family-friendly trail loops around the central portion of the peninsula. More challenging one- to three-mile circuits branch off the central loop. For skilled riders, one trail traverses an elevated ladder bridge, a seventy-five-foot rock crossing and extremely steep hillsides. At the opening celebration with the Southern Appalachian Bicycle Association (SABA) and community supporters, the USFS also announced the completion of picnic tables, exhibits, signage, bathrooms and potable water at the trailhead.

The network of trails, designed and developed with the help of Trail Design Specialists and the International Mountain Biking Trail Care Crew, actually began as a project for Clay County middle school students. In 2001, the school superintendent asked cycling enthusiast and school nurse Joanna Padgett-Atkisson to develop a mountain bike program for its Pathways After-School and Summer Program. After searching local sites, Joanna discussed the project with TVA and the USFS Tusquitee Ranger District. The Forest Service completed its archaeological, environmental and other assessments at an ideal location: the peninsula adjacent to the Jackrabbit Campground. In support of the project, dozens of organizations collaborated with SABA, the USFS and the Clay County Communities Revitalization Association. The funding came from grants, charitable groups, volunteer-donated labor and materials, auctions for two pontoon boats, school fundraising campaigns and other events. Public awareness, community involvement and education in both North Carolina and Georgia became essential. High school seniors gave presentations on building sustainable trails. The local biking club taught students about cycling skills and safety. Not only did regional cyclists come to experience the trail challenges at Jackrabbit, but hikers and cross-country runners did as well.

On September 24, 2016, the Jackrabbit Mountain Bike Trail System hosted more than three hundred cyclists from the Southeast for the sixth annual Chain Buster Endurance Race. Riding laps over an eleven-mile course, individuals and teams competed in three-hour and six-hour endurance events. Over the years, SABA has hosted other regional events as well. "After many years of hard work," said Joanna, "we can now say that it's time to go 'Ride the Rabbit.'"

TVA monitors the ecological health of Chatuge Lake, as it does with all of its other reservoirs. According to TVA, the timing of data collection and the amount of rainfall can affect the assessment results. The agency collects samples at two locations on Chatuge: one on the west side of the lake in the deep waters near the dam and the other on the east side near the mouth of Shooting Creek. The 2012 summary noted conditions similar to previous assessments throughout the 2000s. Fish diversity and totals were lower than expected. Although the sediment quality rated "good"—with no detection of pesticides or concentrations of chromium, copper, nickel or other metals—the dissolved oxygen level rated poor at both collection sites.

During the 1990s, the Chatuge Lake watershed and surrounding areas experienced explosive growth, resulting in a twenty-five-point drop in TVA's ecological health rating. Anglers were catching fewer fish. Historically, the basin had been ecologically rich, providing food sources for the Cherokees and white settlers. More recently, though, officials have documented about seventy-two plant and animal species that are now rare, threatened or endangered. Endemics—like the knotty elimia, Hiwassee crayfish and Hiwassee headwaters crayfish—live nowhere else on the planet. In 2007, the Hiwassee River Watershed Coalition (HRWC) and its partners began monitoring the water quality of the Upper Hiwassee watershed and initiating projects to improve its aquatic habitats.

HRWC members collect water samples on streams and lakes in Clay and Cherokee Counties in North Carolina and Towns and Union Counties in Georgia. In North Carolina, NNF surrounds much of the Hiwassee River Basin. Although TVA and other agencies analyze water samples every few years, HRWC collects monthly samples. Volunteers trained through the Georgia Adopt-a-Stream Program sample the water chemistry at fifty-five stations set up along places, like the Hiwassee, Valley and Nottely Rivers; Fires, Tusquitee and Brasstown Creeks; and tributaries to Nottely and Chatuge Lakes. Members have consistently recorded data collected from eighteen of those sites for the past fifteen years.

Long-term water quality data helps HRWC direct its goals, restoration activities and educational programs. The organization works with other groups, such as the Cherokee and Clay County Soil and Water Conservation Districts, to reduce erosion by stabilizing stream banks and improving riparian habitats. Volunteers remove nonnative, invasive vegetation and plant native trees and shrubs. The roots help filter the soil, reduce runoff and stabilize streambanks. Wide tree canopies absorb sunlight, cast shadows and keep the water cooler. Cooler waters retain more dissolved oxygen and sustain healthier aquatic life. Since 2004, the group has prevented the equivalent of 135 dump truck loads of excess sediment from flowing into the Valley River from unstable stream channels and banks. Since 2011, HRWC has held an annual litter collection campaign on the shores of Chatuge Lake, removing about a ton of trash per year. The organization leads field trips, lectures and workshops. It also offers public education and training to developers, grading contractors, farmers and community leaders to promote a healthy, sustainable water resource in the Hiwassee River Basin.

FIRES CREEK RECREATION AREA

Nine miles east of Murphy, the USFS manages the Fires Creek Recreation Area. A left turn off US 64 follows NC 1302 for four miles to NC 1344. The paved road becomes the graveled FSR 340 and leads into the Fires Creek area. Some historians believe that the creek was named after a family named Fires who lived in the area, but the document *North Carolina 1861—Public Law: Establishment of Clay County* states:

> *Be it enacted by the General Assembly of the State of North Carolina.... That a new county is hereby laid out and established, to be formed out of a portion of Cherokee County, bounded as follows: beginning at the southeast corner of Cherokee County on the Georgia line, thence to run in a northern direction along the top of the Chunckey Gal mountain with the Macon Line, between Shooting creek and Nantahala river to the top of the highest mountain between Fiar's creek and Valley river; thence in a south-west direction along the top of the highest mountain between Fiar's creek and peach tree to the Hiwassee river at the old missionary mill shoal; thence across the Hiwassee River...thence a southward direction...to the Georgia line...thence east with the Georgia line to the beginning.*

The name of Fires Creek seems to be the corrupted version on an old Scottish word meaning "the legally fixed price of corn," according to the *Collins English Dictionary*.

During the 1940s, several large companies harvested Clay County's forests. After Ritter Lumber Company closed its Rainbow Springs sawmill near the current USFS Standing Indian area, it moved to Hayesville. Recently, an Eagle Boy Scout created a covered exhibit near the downtown area, preserving a massive log left by Ritter when it moved to Mountain City, Georgia, in the mid-1960s. The Gennett Lumber Company also logged near Hayesville. However, only the Champion Paper and Fibre Company built a narrow-gauge logging railroad from the mouth of Fires Creek at Hiwassee River to its upper basin. From about 1941 to 1946, Champion transported pulpwood to its paper plant in Canton and sold the lumber to the Boice Hardwood Company. When Champion ended its logging operations near Robbinsville in Graham County, it moved its Shay and Climax engines to its operations at Fires Creek. When Champion left Fires Creek, it sold the engines to the Ely-Thomas Lumber Company in West Virginia. Today, on the Forest Festival Trail at the Cradle of Forestry in America National Historic Site in the Pisgah National Forest sits Champion's historic Climax No. 3, which pulled loads of logs in the Great Smokies and in both Graham and Clay Counties of Nantahala National Forest.

The current USFS road through the Fires Creek area roughly follows Champion's old rail line. The parallel ridges of Tusquitee and Valley River Mountains rise like the grand seating of a massive coliseum above its open-air arena. Fires Creek takes center stage, where horseback riders, campers, hunters, fishermen, day-hikers and backcountry explorers become principal players. Families at the picnic ground wade and tube the creek or hike the

Above: In West Virginia in 1955, Climax No. 3 retired from the logging business. In the 1970s, a Mars Hill College student found it in Michigan, and the USFS bought it. After serving time in NNF and the Smokies, the historic engine is now preserved at the Cradle of Forestry in Pisgah National Forest. *National Forests of North Carolina Historic Photographs, D.H. Ramsey Library, Special Collections, University of North Carolina–Asheville.*

Left: Volunteers with the Unaka chapter of Trout Unlimited help restock Fires Creek with rainbow and brown trout. *Photo by author.*

easy, three-quarter-mile trail to the twenty-five-foot Leatherwood Falls. The Unaka Chapter of Trout Unlimited helps the NCWRC restock the streams with rainbow and brown trout. The rare Appalachian water shrew and the Southern pygmy shrew share this region with the giant hellbender salamander, found only in the eastern United States. Locally known as "waterdogs," some claim that the word *Tusquitee* is Cherokee for "where the waterdogs laughed." Hellbenders also inhabit the deep, dark pools of Tusquitee Creek at the base of the Tusquitee Mountains.

A special beauty, the mountain camellia, found in eight Western North Carolina counties as well as five Piedmont counties, grows in the Fires Creek watershed. According to USFS botanist Gary Kauffman, most populations are relatively small, except for those found in Graham County. The small tree grows in several locations in NNF. It prefers canopy-gap habitats, such as forest edges, streambanks or road banks, where breaks in shade-producing canopies of larger trees allow sunlight to filter down to the understory. The seeds of the mountain camellia may lie dormant in the soil for decades until a nearby wide-branching tree falls, enabling it to germinate. In July 1975, a scrapbook entry by the Nantahala Hiking Club recorded a hike in Ellicott Wilderness by twenty-five people seeking mountain camellias in bloom along the Chattooga River. In March 2016, Brent Martin, Southern Appalachian regional director of The Wilderness Society (TWS), led a day-hike along the Bartram Trail at the Osage Mountain Overlook on NC 106 near Highlands. He urged hikers to return in late June to drive along FSR 79 south of Highlands near the Georgia state line to see mountain camellias in bloom. Another population has been located on FSR 77, south of Brown Gap.

In the Fires Creek basin, TWS sponsors guided hikes to learn more about this cherished species. Designated by the organization as one of North Carolina's "Mountain Treasures," the area is a top priority for wildland protection. Jack Johnston, a local "Stewartia," or mountain camellia expert, often meets TWS hikers at the Leatherwood Picnic Area. Within a short walk on the paved trail, the show begins. During the summer solstice, the mountain camellia opens its palm-sized blossoms of five bright-white petals to exhibit its crowded center of starburst-like filaments topped with golden yellow anthers. Flowers on the same tree may have white, yellow or purple filaments. Its smooth trunk peels back its bark to reveal a grayish-orange inner surface. Although it sports large golden leaves in the fall, the small, sixteen-foot-tall mountain camellia usually goes unnoticed until it blooms in late June. A drive farther along the Forest Service road into the valley provides access to primitive trails and creek crossings that lead to larger stands of mountain camellia. For closer inspections, the forest floor offers handfuls of fallen snowy-white beauties. A rugged bushwhack leads to a specimen listed by the American Forests Champion Trees as the national record for this species.

For trail-hardy, experienced hikers, the twenty-six-mile Rim Trail departs the Leatherwood Picnic Ground to circle Fires Creek Valley along the crests of the Tusquitee and Valley River Mountains. Shorter day-hike trails branch off FSR 340, providing looped options with the Rim Trail. A few road miles beyond the picnic area, the USFS maintains the Bristol Fields Horse Camp with several campsites and a large open area for horse trailers.

BIRDS OF THE NANTAHALA NATIONAL FOREST

By Don Hendershot, naturalist/freelance writer of "The Naturalist's Corner" in Smoky Mountain News *and articles for* Our State, Native American Journal *and others; responsible for USFS bird-point surveys in Nantahala and Pisgah National Forests*

It's late March, about 8:00 a.m.; Horse Branch is gurgling loudly as it snakes its way along the mostly gray and brown forest floor. A few impatient spring beauties and a couple of dazzling white bloodroots have broken through the litter. Suddenly, three loud slurred whistles followed by a musical jumble of notes shatters the quiet stillness of the early spring morning. The Louisiana waterthrush, a diminutive but loud, dapper brown-streaked warbler, is announcing his presence. Louisiana waterthrushes are generally the first of the Neotropical migrants to find their way back to nesting grounds in the Nantahala National Forest. It is believed they evolved such a big voice because they nest and forage along the banks of rushing streams and rivers.

But the floodgates are about to open behind the waterthrush, and by early May, the forest and the skies above it will become a kaleidoscope of color filled with literally millions of Neotropical migrants headed for nesting grounds in North America. Many of those migrants will set up housekeeping in the Nantahala, and those quiet mornings will turn into a cacophony of bird song. More than 150 species of birds utilize the diverse habitats of the Nantahala National Forest.

I have been doing bird point surveys in the Nantahala National Forest as a contractor for the Forest Service since 2007. I annually

The ten-mile-long FSR 340 continues beyond the horse camp to wind deeper into the watershed. A spur route, the gated FSR 340A, becomes the Little Fires Creek Trail and climbs to the rim. Eventually, the main Forest Service road ends at two rugged footpaths that also steeply climb out of the basin to intersect the Rim Trail. The Far Bald Spring Trail takes backcountry hikers to a ridgeline gap between Potrock Bald (5,215 feet) and Tusquitee Bald (5,240 feet). From Potrock Bald, the long-range views span Tusquitee Valley and Chatuge Lake. Some claim that Cherokees sculpted smooth concave, bowl-like depressions near the summit that gave the mountain its name. Did Cherokees carve the basins in the rock for cooking? Did a medicine man use the vessels to concoct herbal remedies?

survey sixty-eight points in the Cheoah and Tusquitee Districts. Protocol calls for spending ten minutes at each point recording all species and individuals heard and/or seen. It is not uncommon to record a dozen or more species numbering more than twenty individuals at a single point. Here is my species list for point 9012 at 4,627 feet elevation on Chunky Gal mountain in the Tusquitee District for May 12, 2016: Canada warbler, ovenbird, black-throated blue warbler, dark-eyed junco, blue jay, ruby-throated hummingbird, eastern towhee, veery, eastern wood-pewee, red-eyed vireo, tufted titmouse, rose-breasted grosbeak and wood thrush—there were twenty-two individual birds.

Birds of the Nantahala and around the world are already being negatively impacted by climate change—a trend that will continue unless the world finds the political and corporate will to address the problem. The forests of the Nantahala with their varied habitats already support "species of concern" like golden-winged warbler, cerulean warbler, whip-poor-will, wood thrush and others. And these intact forests play an important part in the carbon cycle, helping to mitigate climate change.

David Lee was curator of birds at the North Carolina State Museum of Natural Sciences for more than twenty-five years. According to Lee, monitoring efforts by the museum in the old-growth Joyce Kilmer Memorial Forest and adjacent wilderness (areas within the Nantahala National Forest) recorded twenty of the twenty-seven species of wood warblers known to nest in the state. One of the best things we can do for birds is to protect places like the Nantahala National Forest.

Perhaps tribal members collected rainwater in the high-elevation cisterns where spring water was scarce. Hikers have found similar formations near Leatherwood Falls at the picnic area and on the Trail Ridge Trail near the Bristol Horse Camp.

The second footpath at the end of FSR 340 becomes Shinbone Ridge Trail. It leads to County Corners near Weatherman Bald at the northeastern corner of the Rim Trail. A few miles southeast of Weatherman Bald, Tusquitee Bald marks the eastern edge of the Tusquitee Mountains and the highest summit in the national forest district. On Tusquitee Bald, two peaks actually share the same mountain mass. A short ridge joins Tusquitee Peak to its shorter, southern sister-peak, called Signal Bald.

Near the summit of Tusquitee Bald, the Rim Trail intersects the twenty-mile Chunky Gal Trail. Chunky Gal leads about three miles to the former USFS Bob Allison Campground near Tuni Gap. In the late 1880s, the Tusquitee Turnpike left Clay County, followed Tusquitee Creek upstream and crossed Tuni Gap. There it followed the Nantahala River in Macon County. Beyond Big Tuni Creek, the Chunky Gal Trail leaves the Tusquitee Mountains, passes through Perry Gap near Buck Creek and circles Boteler Peak (5,010 feet), the northwestern edge of the long ridge known as Chunky Gal Mountain. After crossing the ridge, the trail dips to Glade Gap to cross US 64 and joins the A.T. near Standing Indian Mountain.

Near County Corners, where Clay, Cherokee and Macon Counties meet, the Shinbone Ridge Trail in Clay County cuts across the rim to enter Macon County. There, the Old Road Gap Trail links the Fires Creek Rim Trail with the USFS network of trails at the Appletree Group Campground, including the Bartram Trail. In 1985, the drug smuggler Andrew Thornton switched his plane to autopilot and parachuted to his death near Knoxville, while his Cessna flew its final voyage toward the Tusquitee District and crashed near County Corners.

In Fires Creek Valley, FSR 340C bears left and leads to a graveled road that ends at a gated footpath. The trail leads to Big Stamp Bald (4,437 feet) on the spine of the Valley River Mountains. Signs remind visitors that Fires Creek is a black bear management area. On the summit, concrete foundations mark the site of the former Big Stamp Fire Tower. Below the summit, a concrete bunker shielded fire wardens from strong thunderstorms.

"You really want me to sit up there all alone in that thing?" recalled Glen Beaver, then a new, young USFS employee. "I have to sit in that tower for days when I've got my eyes on the prettiest little girl in Murphy?" After graduating from Murphy High School, Beaver served in the military police of the National Guard. Later, he worked for the Works Progress Administration (WPA) to stabilize road banks and build recreation areas.

"Then I had to find a steady job," said Beaver. "At that time in Cherokee County, you either went to work at the plant for Levi Straus, like my mother did, or you worked for the Forest Service."

"I had a lot to learn," he continued. "When I started with the USFS, I didn't know the difference between an oak and a poplar tree." After thirty-five years with the USFS, Beaver eventually retired as the seed orchard manager. He was joking about his assignment at the lookout tower, though.

"I actually jumped at the chance to be a fire warden at Big Stamp," he confessed. With a fifteen-pound radio on his back, he hiked a steep five miles to his post. Overlooking the Snowbird, Valley and Tusquitee Mountains, Beaver pined away the time, strumming alone on an old guitar beneath shooting stars by night and clear mountain air by day.

Beaver watched one of the last fires reported from Big Stamp, blazing near Topton, as his replacement flew by. In the 1960s, air surveillance replaced lonely, and not-so-lonely, fire wardens in lookout towers. The orange, inversion smoke, layered near the ground, he said, suggested that the fire was "a hot crown fire, like you get with a fire that's burning in a forest affected by a southern pine beetle infestation."

A view from the Big Stamp Lookout Tower. *National Forests of North Carolina Historic Photographs, D.H. Ramsey Library, Special Collections, University of North Carolina–Asheville.*

FIRE CONTROL OR FIRE MANAGEMENT?

By Chad Boniface, retired USFS resource assistant ranger

Firefighting in the Nantahala National Forest is a common occurrence, mostly in the spring. When the forest burns, many things occur that can be bad, but not always. A lot depends on the fuel consumed in a wildfire. Leaf litter is the most common fuel for most fires. It burns hot and fast. Fuels are categorized into three groups. One-hour fuels, like hardwood leaves, pine needles, tiny twigs and other types of leaf litter, only require one hour of freedom from moisture to dry to combustibility. Ten-hour fuels, such as small branches and other woody materials more than one inch in diameter, require about ten hours of freedom from moisture to dry to a burnable condition. The one-hundred-hour category, which includes branches and trunks greater than five inches in diameter, require four days to dry to combustibility.

In its early days, the Forest Service considered forest fire control as the best science of the day. All forest fires were extinguished as quickly as possible. Indeed, the formation of the USFS was primarily for that purpose. Today, fire prevention and control remain important missions of the agency, preventing loss of property. Almost everyone of any age recognizes Smokey Bear and his message. His story is quite the legend, based on many facts.

The districts in Nantahala have ten to fifteen qualified firefighters who have passed the necessary training and physical requirements to perform these duties. One to two per district usually work as full-time firefighters. Since fires in the East are infrequent and uncommonly occur from natural causes, firefighters perform many other USFS duties during fire-free periods. Across the nation, though, firefighting is a year-round job. Many local firefighters assist with the management of wildfires in other areas as the need arises. Fire management is also the science and application of fire to achieve a desired objective without harming most plants to reduce excess fuels and to enable some species to thrive. Pitch pine, lodgepole pine (western species) and our own table mountain pine would not exist without fire. Fire releases the seeds in the waxy cones to grow and regenerate new seedlings. The growth of many succulent plants is enhanced by burning the understory (plant life on the forest floor). Many species of birds and other wildlife depend on some of the food produced in this scenario. In the East, the native Indian peoples, both food gatherers and hunters, used fire to some degree to produce these effects.

As Big Stamp's last tower operator, Beaver looked out the large windows as the Air Patrol came cruising up the valley "eye level with me." The pilot and his spotter waved as they passed. Ironically, just as they flew by, Beaver spotted smoke. He radioed the pilot.

"L.C., turn around. There's smoke just north of the Andrews-Murphy Airport." Fire management teams responded, and air tankers dropped loads of water. It was contained by daybreak. The USFS retired Big Stamp Lookout Tower soon thereafter and later removed it.

Near the Fires Creek Rim Trail, real estate developers acquired fifty acres near Big Stamp Bald around 2006. If the USFS grants the developers permission to build a road up to a proposed housing project, a portion of the Rim Trail will be rerouted.

The backcountry areas of Fires Creek are rough and remote. For five years, Eric Rudolph, the thirty-six-year-old fugitive wanted by the FBI for bombing an abortion clinic in Alabama and the Olympic Park in Atlanta, hid from authorities in its rugged wilderness. The FBI considered him armed and extremely dangerous and offered a $1 million reward for his arrest. Rudolph's 1996–98 bombings killed two and injured hundreds, placing him on the FBI's Most Wanted List.

The FBI set up its headquarters in Andrews. More than two hundred agents searched NNF, costing the government more than $20 million. The FBI hired local agencies to help with the search. One command post was located at the Appletree Campground. The staff explored forests, old mines, caves and cliffs to no avail. In 2003, a rookie policeman apprehended the fugitive, who was sifting through a dumpster behind the Valley River Shopping Center in Murphy at 3:30 a.m. In a plea bargain, Eric Rudolph pleaded guilty to avoid trial. The courts sentenced him to four consecutive life sentences.

According to his testimony, Rudolph had a summer camp near Murphy and a winter camp on Tarkiln Ridge below Big Peachtree Bald in the Fires Creek area. Some doubt his claims that he survived by eating salamanders, acorns, deer meat and other foods from the forest. He reportedly stole grain from a farmer near the Andrews-Murphy Airport. He also took a truck and one hundred pounds of canned goods from an elderly man's home near Nantahala Lake, leaving $500 on the counter. The FBI believed that he might have received local help.

BUCK CREEK SERPENTINE PINE BARRENS

West of the Macon-Clay County line at the base of Chunky Gal Mountain, Buck Creek ripples through a unique three-hundred-acre ecosystem, known as an ultramafic (rich in iron and magnesium) outcrop barren. Here, in the open woodlands of the largest serpentine barren in the Southern Appalachians, botanists have identified twenty-one state-listed rare plants, including two endemic species. Stunted white oaks mix with pitch pines and prairie grasses normally found in the Midwest. The shallow soils in the Buck Creek barrens have a higher pH level and different clay content than other soils in the Nantahala region and help determine what plant communities thrive here.

Uncommon in the Southern Blue Ridge, dunite, olivine and other minerals in the Buck Creek area contribute to its soil composition. Millions of years ago, when oceanic tectonic plates collided, driving plate edges beneath the earth's crust, water seeped into sediments and rock fractures on the ocean floor. Hydrothermal metamorphosis changed or replaced some of the rock's minerals into serpentine minerals, a group of igneous materials consisting mostly of hydrous magnesium sulfate. Like snakeskin, these serpentine minerals display a slick, oily greenish-brown pattern. Geologic time and changes eventually exposed olivine and other serpentine minerals on the earth's surface. Fine-grained serpentine rocks eventually weather and form soils with low levels of calcium and nitrogen; high levels of magnesium; and increased concentration of nickel, chromium and other heavy metals. Only plants that have adapted to these extreme conditions can survive in the serpentine barrens.

Since the early 1980s, officials have designated 103 Buck Creek acres, now registered with the North Carolina Natural Heritage program, as a special use area. In 1995, the USFS began actively restoring special plant communities in the serpentine barrens (3,500–4,000 feet) with prescribed burns to limit competitive vegetation. To assess the effectiveness of the fire management plans, the North Carolina Vegetative Survey established seven research plots. In 2007, University of North Carolina–Chapel Hill biology student Elizabeth Marx evaluated the interactions between fire and flora on six of those study areas. She found that prescribed burning reduced the canopy cover, providing the sunlit environment preferred by many of the rare and endemic serpentine species. On the most frequently burned plots, six rare grasses grew significantly better than on other plots that maintained a tree cover greater than 50 percent. Western Carolina

University groups, plant societies, scientists and others come to Buck Creek to study this unique ecosystem.

Amateur and experienced prospectors screen-sift and surface-collect fluorite; tourmaline; blue, gray and pink corundum; a little gold; a lot of garnets; and other rocks and minerals. They stumble up rocky gullies, over gravel bars of old tailings and down creek beds looking for treasures. They find large garnet-studded rocks, pocket-sized corundum chunks and bits of mineral-rich sand and gravel. Beyond an old lake, now a species-rich bog area, a gated roadway leads about one-half mile to the site of the old Herbert Mine. Although some mining began after 1874, when C.D. Smith discovered corundum in the Buck Creek area, the North Carolina Corundum Company set up a mining camp and operated the Herbert Mine from 1903 to 1906. A trail off the old roadbed leads to nice rockhound sites along Barnard Creek.

By the early 1900s, Andrews Manufacturing Company (directing operations of the Andrews Lumber Company and later becoming the Andrews Hardwood Company) controlled nearly eight thousand acres of hardwood and hemlock in the Nantahala region near the Macon-Clay County line. The general manager of the Andrews company, William T. Latham, established a large logging camp near the headwaters of Buck Creek. He eventually built a spur rail line north along Buck Creek to its confluence with Nantahala River. There, it joined rail lines headed northwest past Aquone to a large sawmill at Andrews near the Murphy Branch rail line. Later, Latham helped form the Badgett and Latham Company, directing logging projects on lands owned by Andrews Hardwood. West of Buck Creek, the company extended a logging rail line south of Aquone through Tuni Gap in the Tusquitee Mountains, past the closed Bob Allison Picnic Area and along Big Tuni Creek to its juncture with Tusquitee Creek.

Near the site of the former Corundum Mill in Buck Creek, Latham managed the logging camp until 1924. By 1931, he had built a mountain lodge at Buck Creek within view of Doe Knob. Faded scrapbook photos show Latham family members standing on the upper and lower balconies of the old Corundum Mill's office building surveying the property. Latham added a fishing lake and a barn for sheep. He added the Upper Cabin, the Brown Cabin and the Bobby House. David Latham split 1,500 clapboards to cover Bobby's roof. Railroad ties became fence posts around its five-acre plot. By 1934, a fire had consumed the main lodge.

The next spring, Latham began constructing a larger mountain retreat to accommodate forty guests. After an eighteen-month delay, he opened "a new ranch house...the finest lodge in Carolina...with ten miles of trout streams on the vast estate," reported a 1938 newspaper. On the west side, visitors entered the twenty-eight-room lodge. A white oak staircase led to the upper-level bedrooms with shower baths. On the main level, hemlock tree trunks supported exposed beams and a twelve-foot ceiling. Sunlight penetrated the lounge through nearly two dozen huge windows. Cherry tree branches supported a solid cherry tabletop. Cherry bookcases lined the wall beside a wide fireplace. During its grand opening year, the lodge hosted its first major social event, the wedding of David Latham and Nadine Johnson.

In the large dining room, a buffet served butter and milk from the estate's Jersey herd, fruits and vegetables from its gardens, hickory-smoked pork from its farmyard and fresh fish from its streams. Latham dammed Barnard Creek, creating trout rearing pools to restock Buck Creek. A flagstone-terraced patio invited guests to enjoy the mountain views. Outside, flowers lined the pathway to the Upper Cabin on the hill overlooking the lodge. According to the scrapbook, one of the former logging cabins protected "ewes who were ready to lamb. But, wildcats soon caught on and got the lambs through the windows." Other buildings included a servant's quarters and a garage. A shed held two generators providing electrical power. For entertainment, Latham built horse stables, a football field and a swimming pool. He also planned croquet fields, shuffleboards and tennis courts.

During the 1930s and '40s, Latham's Buck Creek Ranch brochure promoted the beauty of the Nantahala region:

> *Come to Buck Creek midway between Franklin and Hayesville for a delightful vacation...a restful, comfortable ranch at 3500 feet in the midst of the Nantahala National Forest, a mountain wilderness of rugged beauty. Standing Indian, one of the most picturesque mountains in Western North Carolina, is nine miles from the ranch and is a delightful place for picnic parties. On the crest of the mountain is a government fire tower. Wayah Bald, another of the lofty peaks, is within a short drive over winding forest roads. On this peak stands one of the most imposing stone fire towers in the forest. National forest trails lead from the ranch to and along the high tops of the mountains through unbroken forests for miles and are ideal for horseback riding or hiking.*

In the early 1950s, Latham's wife died, and he left Western North Carolina. For a brief time, others managed the Buck Creek resort as the Silver Spur Ranch. Soon, Latham sold the property to a Catholic couple,

George and Virginia Wright, who eventually deeded it to the Glenmary Catholic Missioners of Rural America. Between May 1961 and June 1964, the Glenmarians acquired the former Latham lodge and about 475 acres of land in three separate tracts.

Headquartered in Cincinnati, Ohio, the Glenmary Home Missioners, now celebrating more than seventy-five years of home mission ministry, provide spiritual, social and economic outreach at fifty mission sites in rural communities of the South and Southwest, as well as the Appalachian Mountains. Founded in 1939 by Father William Howard Bishop, this Catholic society "is the only religious community devoted exclusively to serving the spiritually and materially poor in US home missions," stated a 2007 edition of the *Glenmary Challenge*. In 2003, Father Jim Wilmes—a Glenmarian for forty-six years who served in Murphy, Andrews and Buck Creek—died in Robbinsville shoveling snow outside the small room where he lived in the Prince of Peace Church. His obituary paid tribute to his service and stated, "Glenmary priests were the first Catholic clergy to bring Catholicism to rural North Carolina, holding tent revivals during the summer and visiting the people of the area door to door."

Protestants had organized churches in Clay County in the 1940s. By the 1950s, Glenmary nuns of the Roman Catholic Church had purchased the old Hayesville Motel for a convent and chapel. The Catholics searched for an ideal location for a training center for newly ordained priests who would serve in local missions. Before serving unfamiliar societies of other nations, foreign missionaries would study that region's culture, habits, interests and needs. Likewise, the Catholic priests and brothers of Glenmary prepared their missioners by educating them about the sociocultural attributes of rural America. Surrounded by Nantahala National Forest, remote Buck Creek and lodge provided an idyllic setting for the Pius XII Pastoral Center.

On the north wall of the old Buck Creek Lodge, the Glenmary Artisan Building Crew removed two levels of wooden decks on an A-frame façade to build the Pastoral Center Chapel perpendicular to the main lodge building. Two high foundation walls of native stones held the steep-roofed structure above the sloping terrain. Accentuating a high ceiling, wood panels placed at forty-five-degree angles framed a wall of glass in the front of the sanctuary. A floor-to-ceiling wooden cross braced the large window panes. Inspiring peace and reflection, grand views of the natural world enhanced the religious experience and the offering of Mass. On August 28, 1963, His Excellency, Bishop Vincent Waters of Raleigh, came to Buck Creek. "Bishop will Bless Glenmary Center," announced the *North Carolina Catholic*:

A narrow graveled road winds through swamp alder, birch and whispering spruce. It rattles over loose board bridges and glances sideways into rushing white water with here and there a deep clear pool. Big-eyed deer stand silently in the shadows. Night time sends the terror of a bobcat's call into scampering furry creatures on the forest floor. This is Buck Creek, North Carolina, home of the Glenmary's Pastoral Training Center.

The Glenmaries hired Carl Lyvers of Hayesville as supervisor of the training center grounds. At his home on Tusquitee Creek in November 2016, Carl secured his vise holding a chair, awaiting repair in his woodworking shop. Tony, one of his ten children, moved tools from chairs to make room for an interviewer to sit.

"I love to talk about Buck Creek," said the soft-spoken, pleasant eighty-year-old. "My family lived in the old Latham three-bedroom house for ten years. It became home to me. It still had the original handmade furniture in it."

Pius XII Pastoral Center of the Glenmary Home Missioners at Buck Creek. *Glenmary Missioners, Cincinnati, Ohio.*

"I practically grew up there," echoed Tony, now in his sixties. "I was five when we moved there."

In charge of maintenance, machinery and security, Carl restored the old Bobby House and other buildings. He repaired doors, windows, walkways and furnaces. He helped build the chapel, the entrance gate and a new half-acre trout lake. During the summer of '58 or '59, campers at Glenmary's Caro Lina Camp for Boys and others helped Carl replace the swimming pool built by the Lathams in the 1930s.

"I think it was an Olympic-sized swimming pool," said Tony, who joined the young campers in their activities. "At least it seemed that big to me! Can you imagine growing up here in the mountains of Clay County and having free access to a deluxe swimming pool all summer!" A rock wall and bathhouse bordered the pool beside the diving board. Remnants of the old pool remain in an open field.

Motorcoach-sized buses owned by the Glenmarians transported young campers four hundred miles from their headquarters in Cincinnati to Buck Creek. Along the way, chaperones treated the thirteen- to nineteen-year-old boys to tourist stops at Gatlinburg and Cherokee. The camp was a "unique experience for any boy who has expressed a desire for religious life," promoted the Catholic society's brochure, "a ten-day vocation camp for boys interested in the priesthood or brotherhood." In Cincinnati, four days of classes prepared the young men for the six-day outdoor retreat and religious discussions with the Glenmary Fathers at Buck Creek.

Recreation included fishing, swimming, Junior Olympic games and nature study. Tony spent many summers with the young campers, leading them on hikes "all over the mountains at Buck Creek." He took them to his favorite rockhounding sites.

"We collected corundum like these," he said pulling out his bucketful collection. In addition to soapstone, mica and other minerals, they occasionally found flint stone arrowheads that Cherokees had received in trade with western tribes. "We usually found gray, blue and pink corundum at Buck Creek," Tony explained. Years later, Tony found a large ruby-red chunk that he made into a necklace for his wife. On a trip to California in 1992, the couple lost the treasure in debris after an earthquake.

By 1968, the Pius XII Pastoral Center had become too expensive to maintain. The Catholic society sold the property to five investors. After improving the road and building a new bridge, Carl Lyvers moved back to his home on Tusquitee Creek. By 1973, two of the investors kept part of the Buck Creek property to build private homes. They sold the rest, including

BUCK CREEK RANCH

In the Heart of The Nantahala National Forest

Cool, Quiet, Restful

About two miles north of U. S. 65
Midway between Franklin and Hayesville
Mr. and Mrs. W. T. LATHAM, Owners
RAINBOW SPRINGS - NORTH CAROLINA

Left: The brochure for W.T. Latham's Buck Creek Ranch asked guests to notify the ranch three days in advance of their arrival so that staff could meet them at their entrance road on US 64. Guests could send telegrams from Franklin by homing pigeons "that are raised and trained at the ranch." *Latham Family Scrapbook, Glenmary Missioners, Cincinnati, Ohio.*

Below: Ranger William Nothstein begins a hike on the Appalachian Trail. *Macon County Historical Society and Museum.*

the lodge, to a real estate developer in Macon County. Bankruptcy, however, prevented Leo Furlong from building his proposed mountain resort. In 1978, the USFS acquired the Buck Creek property from a mortgage company.

Described by the Forest Service as a "highly scenic area of Nantahala National Forest," officials discussed turning the lodge into a USFS administrative building or training center. Repair costs, though, seemed prohibitive. The building had no insulation. More than $200,000 would be required to repair structural decay, improve safety risks, upgrade eroded roadways and replace outdated wiring and plumbing. Ongoing maintenance would be expensive. Some suggested demolition. Theft and vandalism of valuable wormy chestnut had already become a problem. More than eight break-ins resulted in thousands of dollars of materials lost, stated one newspaper. "One of the old buildings was stripped of everything but the stud walls and roof in one night. Plumbing, windows, doors, plywood walls, and other things were taken." Thankfully, the article continued, the Forest Service had already given the antique, handmade furniture to Southwestern Technical College in Sylva.

While the USFS heard comments and opposition at public meetings about demolishing the historic structure, the former Buck Creek Lodge and all its outbuildings mysteriously burned to the ground overnight on March 10, 1981. After years of mining and logging operations and after providing a mountain home for a logger's lodge, a dude ranch, a Catholic priests' retreat and a boys' summer camp, the Buck Creek Serpentine Pine Barrens are now protected by the USFS.

William Nothstein, thirty-four-year veteran of the USFS, wrote:

> *As the resources and facilities of the Cheoah Ranger District* [as well as the Tusquitee and Nantahala Districts] *became better known and available, they attracted more people. Each group wanted preference and priority for its favored activity. Some conflicts developed. In time,* [the controversies] *may be viewed as fortunate since they should lead to the best possible solutions for the benefits of most people. Be happy with freedom, space, clean air, clean water and a respect for nature!*

Some people have spent thirty years, working forty hours a week, for the national forest system—and not just to earn an income, maintain a dedicated work ethic or uphold a solid sense of commitment. The forest became a part of their identity, their soul. After thirty years of service, Marshall McClung retired from Nantahala National Forest and now works as a firefighter with the state forest service. The year 2017 will be the fiftieth year of his life spent fighting wildfires. Hoot Gibbs has also spent most of his life in the forest; now retired from the USFS, he packs in as much weight as the donkey, Alf, to maintain Nantahala's trails. Joe Bonnette, retired district ranger, became the director of the Partners of Joyce Kilmer–Slickrock Wilderness, a volunteer organization assisting the USFS Cheoah District with upgrades, restoration, preservation and other projects. Many others have retired from the Forest Service but not from the forest.

"We love what we do," said retired district ranger Lewis Kearney. "We love the forest."

"We are stewards of a great national trust," said another district ranger.

The public, too, loves a national forest. People come for inspiration and peace among cathedral-like columns of massive trees; for wide, distant views from granite domes; and for exhilarating adventures beneath the noonday sun in a deep gorge. Environmentalists and scientists come to protect plant communities and ecosystems, change the direction of a skyway or preserve the quality of watersheds and trout waters. Researchers and archaeologists retrace Cherokee footpaths and excavate detainment forts. Some come to follow a trail or river, identify a bird or flower, find a rock or gem or catch a fish or two. They straddle bicycles, horses and ATVs. Some hunt wildlife; others erect poles for a safe glide across a highway. Volunteers clear a bald to preserve flame azaleas; collect trash along roadways, streambanks and lakeshores; help restock trout streams; and restore trails, overlooks and recreational areas.

Historians study the region's past—its former property owners and the USFS acquisition of the acreage; the railroads, steam engines and logging operations; the construction of dams and reservoirs; and the history of mountain lodges, some still open for guests and others now faded memories.

The 500,000 acres of Nantahala National Forest, with the USFS as its resource manager, entertains the interests of a wide range of passions and personalities. The goal, perhaps, for both the manager and the users, is to protect, enjoy and sustain the forests and its resources so the next generation can love it the way we do.

APPENDIX

CHECKLIST OF PLANTS AND ANIMALS MENTIONED IN THE TEXT

Trees and Shrubs

Alternate-leaf dogwood, *Cornus alternifolia*
American basswood, *Tilia americana*
American beech, *Fagus grandifolia*
American chestnut, *Castanea dentata*
Black cherry, *Prunus serotina*
Black walnut, *Juglans nigra*
Butternut (white walnut), *Juglans cinerea*
Carolina silverbell, *Halesia carolina*
Catawba rhododendron (purple rhododendron), *Rhododendron catawbiense*
Clammy azaleas, *Rhododendron viscosum*
Common silverbell, *Halesia tetraptera*
Cucumber magnolia, *Magnolia acuminata*
Dog-hobble, *Leucothoe fontanesiana*
Douglas fir, *Pseudotsuga menziesii*
Eastern hemlock, *Tsuga canadensis*
Flame azalea, *Rhododendron calendulaceum*
Flowering dogwood, *Cornus florida*
Franklin tree, *Franklinia alatamaha*
Fraser fir, *Abies fraseri*
Fraser magnolia, *Magnolia fraseri*

Fringetree, *Chionanthus virginicus*
Hartweg's locust, *Robinia viscosa var. hartwegii*
Ironwood, *Ostrya virginiana*
Minniebush, *Minziesia pilosa*
Mountain camellia, *Stewartia ovata*
Mountain laurel, *Kalmia latifolia*
Northern red oak, *Quercus rubra*
Norway spruce, *Picea abies*
Pawpaw, *Asimina triloba*
Pinkshell azalea, *Rhododendron vaseyi*
Red maple, *Acer rubrum*
Red spruce, *Picea rubens*
Sandmyrtle, *Leiophyllum buxifolium*
Shortleaf pine, *Pinus echinata*
Smooth (white azalea), *Rhododendron arborescens*
Southern magnolia, *Magnolia grandiflora*
Southern red oak, *Quercus falcata*
Striped maple, *Acer pensylvanicum*
Sugar maple, *Acer saccharum*
Sweet or black birch, *Betula lenta*
Sycamore, *Platanus occidentalis*
Table mountain pine, *Pinus pungens*
Tulip poplar, *Liriodendron tulipifera*
Virginia pine, *Pinus virginiana*
Virginia spiraea, *Spiraea virginiana*
Western hemlock, *Tsuga heteropylla*
White oak, *Quercus alba*
White pine, *Pinus strobus*
Witch-hazel, *Hamamelis virginiana*
Yellow birch, *Betula alleghaniensis*
Yellow buckeye, *Aesculus octandra*

WILDFLOWERS

Beebalm, *Monarda didyma*
Bishop's cap, *Mitella diphylla*
Black cohosh, *Cimicifuga racemosa*

Bleeding heart, *Dicentra eximia*
Bloodroot, *Sanguinaria canadensis*
Blue cohosh, *Caulophyllum thalictoides*
Bog goldenrod, *Solidago uliginosa*
Bowman's root, *Porteranthus trifoliatus*
Buckley's St. John's-wort, *Hypericum buckleyi*
Cardinal flower, *Lobelia cardinalis*
Columbine, *Aquilegia canadensis*
Coneflower, *Rudbeckia laciniata var. humilis*
Crested dwarf iris, *Iris cristata*
Curtis' goldenrod, *Solidago curtisii*
Cuthbert's turtlehead, *Chelone cuthbertii*
Divided-leaf ragwort, *Senecio millefolium*
Dutchman's breeches, *Dicentra cucullaria*
Fire pink, *Silene virginica*
Foamflower, *Tiarella cordifolia*
Ginseng, *Panax quinquefolius*
Goatsbeard, *Aruncus dioicus*
Granite dome bluets, *Houstonia longifolia var. glabra*
Grass pink, *Calopogon tuberosus*
Jack-in-the-pulpit, *Arisaema triphyllum*
Japanese rose, *Kerria japonica*
Kidney-leaved grass-of-Parnassus, *Parnassia asarifolia*
Large-flowered bellwort, *Uvularia grandiflora*
Large purple-fringed orchid, *Platanthera grandiflora*
Purple-flowering raspberry, *Rubus odoratus*
Quill fameflower, *Phemeranthus teretifolius*
Showy orchis, *Galearis spectabilis*
Speckled wood lily, *Clintonia umbellulata*
Spring-beauty, *Claytonia caroliniana*
Squirrel corn, *Dicentra canadensis*
Star chickweed, *Stellaria pubera*
Three-lobed black-eyed Susan, *Rudbeckia triloba*
Turk's cap lily, *Lilium superbum*
Umbrella leaf, *Diphylleia cymosa*
Wake robin, *Trillium erectum*
Whiteleaf sunflower, *Helianthus glaucophyllus*
White trillium, *Trillium grandifolia*
Yellow lady's slipper, *Cypripedium pubescens*

Grasses, Sedges, Rushes and More

Biltmore sedge, *Carex biltmoreana*
Deerhair bulrush, *Scirpus caespitosus*

Insects, Spiders, Butterflies and More

Lost Nantahala cave spider, *Nesticus cooperi*
Pipevine swallowtail, *Battus philenor*
Southern pine beetle, *Dendroctonus frontalis*

Ferns and Fern Allies, Mosses and Liverworts, Lichens and More

Appalachian clubmoss, *Huperzia appalachiana*
Appalachian filmy-fern (Appalachian bristle fern), *Trichomanes boschianum*
Glade fern, *Diplazium pycnocarpon*
Goldie's fern, *Dryopteris goldiana*
Maidenhair fern, *Adiantum pedatum*
Northern beech fern, *Phegopteris connectilis*
Rock clubmoss, *Huperzia porophila*
Rock gnome lichen, *Gymnoderma lineare*
Twisted-hair spikemoss, *Selaginella tortipila*
West Indian dwarf polypody fern, *Grammitis nimbata*

Fish, Crustaceans, Amphibians and Other Aquatic Species

Brook trout, *Salvelinus fontinalis*
Brown trout, *Salmo trutta*
Citico darters, *Etheostoma sitikuense*
Eastern hellbender salamander, *Cryptobranchus alleganiensis*
Hiwassee crayfish, *Cambarus hiwasseensis*
Hiwassee headwaters crayfish, *Cambarus parrishi*

Junaluska salamander, *Eurycea junaluska*
Knotty elimia, *Elimia interrupta*
Little Tennessee crayfish, *Cambarus georgiae*
Noonday globe snail, *Petera clarkia Nantahala*
Rainbow trout, *Oncorhynchus mykiss*
Sicklefin redhorse, *Moxostoma sp.*
Smoky madtom, *Noturus baileyi*

Birds

Barred owl, *Strix varia*
Belted kingfisher, *Ceryle alcyon*
Black-throated blue warbler, *Setophaga caerulescens*
Blue-gray gnatcatcher, *Polioptila melanura*
Eastern towhee, *Pipilo erythrophthalmus*
Gray catbird, *Dumetella carolinensis*
Great blue heron, *Ardea herodias*
Mallard, *Anas platyrhynchos*
Red-bellied woodpecker, *Melanerpes carolinus*
Red-tailed hawk, *Buteo jamaicensis*
Ruffed grouse, *Bonasa umbellus*
Veery, *Catharus fuscescens*
Wild turkey, *Meleagris gallopavo*
Yellow-billed cuckoo, *Coccyzus ameicanus*

Mammals

American black bear, *Ursus americanus*
American buffalo, *Bison bison*
Carolina northern flying squirrel, *Glaucomys sabrinus coloratus*
European wild boar, *Sus scrofa*
Southern flying squirrel, *Glaucomys Volans*
Southern pygmy shrew, *Sorex hoyi winnemana*
Southern water shrew, *Sorex palustris punctulatus*
White-tailed deer, *Odocoileus virginianus*

BIBLIOGRAPHY

Adams, Allison O. "A River Runs Through It." *Emory* (1996). http://www.emory.edu/EMORY_MAGAZINE/winter98/kennedy.html.

Adams, Kevin. *North Carolina Waterfalls.* Winston-Salem, NC: John F. Blair, Publishers, 2016.

Ager, John Curtis. *We Plow God's Fields: The Life of James G.K. McClure.* Boone, NC: Appalachian Consortium Press, 1991.

Alexander, Phil. "Groups Debate Use of Roy Taylor Forest." *Asheville Citizen-Times*, May 14, 1993.

———. "Volunteer Group Looks at Using Roy Taylor Forest." *Asheville Citizen-Times*, May 25, 1993.

Alley, Felix E. *Random Thoughts and the Musings of a Mountaineer.* Salisbury, NC: Rowan Printing Company, 1941.

American Whitewater. "Chattooga Project." 1999–2017. http://www.americanwhitewater.org/content/Project/view/id/chattooga.

Ammons, Lindy, and Lisha Ammons, Graham County historians. Personal interview, April 17, 2016.

Arthur, John Preston. *Western North Carolina: A History from 1730 to 1913.* Asheville, NC: Edward Buncombe Chapter of the Daughters of the American Revolution, 1914.

Ashe, Alicia. "Forest Service Set to Upgrade Nantahala River Trails, Access." *Franklin Press*, December 7, 1992.

Asheville and Buncombe County. "Nimrod and Nancy Jarrett." March 3, 2010. http://ashevilleandbuncombecounty.blogspot.com/2010/03/nimrod-and-nancy-avaline-jarrett.html.

Asheville Citizen-Times. "Acreage in Six Mountain Counties to Be Added to Nantahala Forest." March 25, 1934.

———. "Crouch Property May Be Added to Nantahala National Forest." August 2, 1990.

———. "Exercises to Be Held on Wayah Bald on Sept. 6." August 29, 1937.

———. "Extension of Parkway Favored." June 23, 1961.

———. "First Ranger Station Still Stands." November 13, 1966.

———. "446,650 Acres in Nantahala." June 28, 1959.

———. "Mountain in Public Domain." March 5, 1973.

———. "Nantahala: High Peaks, Beautiful Narrow Valleys." June 24, 1962.

———. "Nantahala Is Place of Charm." June 17, 1940.

———. "Nantahala Package Go-Ahead Report." March 20, 1977.

———. "NC's Wayah Crest and Wayah Bald." August 28, 1949.

———. "New Land Will Go to Nantahala Forest Preserve." March 30, 1934.

———. "Watershed Protection and Management Are Emphasized in Nantahala National Forest." June 28, 1959.

Averell, Graham. "Mapping the Secrets of Big Pisgah and Bonas Defeat." Blue Ridge Outdoors, April 6, 2010 http://www.blueridgeoutdoors.com/go-outside/hiking/mapping-the-secrets.

Barcott, Bruce. "Eric Rudolph Slept Here." *Outside* (September 1, 2003).

Barr, Peter. *Hiking North Carolina's Lookout Towers.* Winston-Salem, NC: John F. Blair Publisher, 2008.

Bartram Society. *Bartram Heritage.* Montgomery, AL: Bartram Trail Conference, 1979.

Beaver, Glenn, retired USFS. Personal Interview, Murphy, NC, October 28, 2016.

Boniface, Chad, retired USFS. Three guided tours of Nantahala Ranger District, interview and e-mail conversations, 2016–17.

Boyd, Brian. *The Highlands-Cashiers Outdoors Companion.* Tallulah Falls, GA: Fern Creek Press, 2011.

Brammer, Jack, and Michael York. "Cocaine Discovery linked to Thornton." *Lexington Herald-Leader*, September 11, 1985.

Brown, Scooter, retired USFS. Personal interview, Robbinsville, NC, summer 2016.

Buck Creek Ranch. Brochure, no date.

Byrd, Richard. "Land of the Noonday Sun." *North Georgia Journal* (Winter 1995).

Cantrell, H. Furman. "Native Azaleas on Wayah Bald." *Journal American Rhododendron Society* 34, no. 4 (Fall 1980).

Cargas, Harry J. *I Lay Down My Life: Biography of Joyce Kilmer.* Boston, MA: Daughters of St. Paul, 1964.

Carter, Mason C., Robert C. Kellison and R. Scott Wallinger. *Forestry in the U.S. South: A History.* Baton Rouge: Louisiana State University Press, 2015.

Cartwright, James B. "Nantahala National Forest Has Great Wealth of Shrubs and Flowers." *Asheville Citizen-Times*, June 23, 1935.

Chambers, Isabel, and Overton Chambers. *Remembering Highlands.* Charleston, SC: The History Press, 2009.

Cherokee Phoenix. "Eastern Band Buys Cowee Mound." January 8, 2007. http://www.cherokeephoenix.org/1793/Article.aspx.

Cochran, Weimer. *Nantahala People and Events.* Booklet. Berea, KY: Berea College Appalachian Center, 1987.

Cook, Chad, USFS Fire Management Officer, Cheoah/Tusquitee Ranger District. Guided tour of Beech Creek Genetic Resources Management Area, October 2016.

Corey, Jane. *Exploring the Waterfalls of North Carolina.* Chapel Hill, NC: Provincial Press, 1991.

Corsair, Gary. "Joyce Kilmer Scarred." *Graham Star*, December 15, 2016.

———. "Memorial Planned at Crash Site." *Graham Star*, October 15, 2015.

———. "Tribute to Final Flight." *Graham Star*, October 22, 2015.

Costa, James T. *Highlands Botanical Garden: A Naturalist's Guide.* Highlands, NC: Highlands Biological Foundation Inc., 2012.

Cross, Arthur. "Nantahala." http://www.visitnantahalanc.com.

De Hart, Allen. *North Carolina Hiking Trails.* Boston: MA: Appalachian Mountain Club Books, 1988.

Denton, David Morris. *The Denton Experience.* Baltimore, MD: Publish America, 2004.

Denton, Sally. *The Bluegrass Conspiracy: An Inside Story of Power, Greed, Drugs and Murder.* Lincoln, NE: published by the author, 2016.

Duncan, Barbara R., and Brett H. Riggs. *Cherokee Heritage Trails Guidebook.* Chapel Hill: University of North Carolina Press, 2003.

Early, Lawrence, and Curtis Wooten. "Taylor Crockett Remembers." *Wildlife in North Carolina* (February 1985).

Ellison, George. *Literary Excursions in the Southern Highlands.* Charleston, SC: The History Press, 2016.

———. "Looking for Tsali." *Asheville Citizen-Times*, 1991.

———. *Mountain Passages: Natural and Cultural History of Western North Carolina and the Great Smoky Mountains.* Charleston, SC: The History Press, 2005.

———. Personal interview, Bryson City, NC.

———. "Purchase Would Help Preserve Whiteside Mountain Area." *Asheville Citizen-Times*, n.d.

———. "Zahner's Special Affection for Highlands." *Smoky Mountain News*, September 9, 2009. http://smokymountainnews.com/news/item/1740-zahner%E2%80%99s-special-affection-for-highlands.

Ellison, George, and Lance Holland. *Nantahala River: Land of the Noonday Sun.* Video. Bryson City, NC: Lance Holland, 1998.

Ellison, Quintin. "The Great Wasilik Poplar's Time at the Top May Be Numbered." *Smoky Mountain Journal,* May 4, 2011.

———. "New Shelter Graces Wayah Bald." *Smoky Mountain News,* June 6, 2007.

English, Donald B.K., and Michael A. Tarrant. "A Crowding-Based Model of Social Carrying Capacity: Application for Whitewater Boating Use." 1996. http://www.srs.fs.usda.gov/recreation/nantahalappr.pdf.

Federal Bureau of Investigation. "The Pursuit and Capture of Eric Rudolph." May 5, 2005. https://www.fbi.gov/news/stories/2005/may/swecker_051605.

Fetters, Thomas. *Logging Railroads of the Blue Ridge and Smoky Mountains.* Vol. 2. Hillsboro, OR: TimberTimes Press, 2010.

Fischer, Kelly. *Playboating the Nantahala River.* Almond, NC: Milestone Press, 1988.

Forest History Society. "Weeks Law Purchase Units, 1911–1932." Chronology of National Forests established under the Weeks Act. www.foresthistory.org.

Frank, Bernard. *Our National Forests.* Norman: University of Oklahoma Press, 1955.

Frankenberg, Dirk. *Exploring North Carolina's Natural Areas: Peaks, Nature Preserves, and Hiking Trails.* Chapel Hill: University of North Carolina Press, 2000.

Franklin Press. "Forest Service Constructs and Names Dr. Morgan Trail." October 15, 1979.

———. "Opal's Effect on Forest Widespread." November 8, 1995.

———. "Senior Citizens Mean a Lot to Local Camping Areas." March 12, 1981.

Gaddy, L.L. *A Naturalist's Guide to the Southern Blue Ridge Front.* Columbia: University of South Carolina Press, 2000.

Garren, Laura Ann. *The Chattooga River: A Natural and Cultural History.* Charleston, SC: The History Press, 2013.

George, Michael. *Southern Railway's Murphy Branch.* Collegedale, TN: College Press, 1996.

"Georgia Parkway Extension: The Unbuilt Blue Ridge Parkway" / "Georgia Extension: Panthertown." A Digital Public History Project. University of North Carolina–Chapel Hill, Fall 2013.

Glenmary Challenge. "Western North Carolina: Glenmary's Legacy: 16 Missions Established in 12 Counties" (Winter 2009).

Gordon, Samuel G. "Corundum Hill (Franklin), Macon County, North Carolina." *Journal Mineralogical Society of Natural Sciences of Philadelphia.* http://www.minsocam.org/ammin/AM7/AM7_189.pdf.

Graham, Eddie, USFS volunteer. Guided tour of Panther Top Fire Tower, Tusquitee District, October 2016.

Graham County Centennial. Booklet. Robbinsville, NC: Graham County Centennial Inc., 1972.

Hainge, Kim. "Airmen of John's Knob Remembered." *Graham Star*, October 29, 2015.

———. "Protecting Hooper Bald's Amazing Azaleas." *Graham Star*, August 4, 2016.

Hairr, John. *North Carolina Rivers.* Charleston, SC: The History Press, 2007.

Harget, Gil Forest. *Great Adventures in the Southern Appalachians.* Winston-Salem, NC: John F. Blair, Publisher, 1994.

Harper, Francis. *The Travels of William Bartram, Naturalist's Edition.* Athens: University of Georgia Press, 1998.

Harshaw, Lou. *The Rubies of Cowee Valley.* Booklet. Asheville, NC: Hexagon Company, 1973.

Hatter, Ila, naturalist, educator and guide. Guided hike in Joyce Kilmer Memorial Forest, July 2016.

Hayler, Nicole, ed. *Sound Wormy.* Athens: University of Georgia Press, 2002.

Hendershot, Don. "Cherokee Fish Weirs." *Smoky Mountain Living*, August 1, 2013.

History.net. "U.S. Torpedo Troubles during World War II." June 12, 2006. http://www.historynet.com/us-torpedo-troubles-during-world-war-ii.htm.

Hiwassee River Watershed Coalition. 2004–13. www.hrwc.net.

Holland, Lance. *Fontana: A Pocket History of Appalachia.* Robbinsville, NC: Appalachian History Series, 2001.

Homan, Tim. *Hiking Trails of the Joyce Kilmer–Slickrock & Citico Creek Wilderness.* Atlanta, GA: Peachtree Publishers, 2008.

———. *Hiking Trails of the Southern Nantahala Wilderness, Ellicott Wilderness, Chattooga National Wild and Scenic River.* Atlanta, GA: Peachtree Publishers, 2002.

Howard, Jason. "Appalachia Burning." *New York Times*, November 30, 2016.

Hoyle, Zoe. "White Pines, Hemlocks and Sunlight." *CompassLive Newsletter*, December 29, 2016.

Ianniello, Jan, and Steve Ianniello. *Guide to Day Hikes and Fly Fishing and Mountain Biking.* Highlands, NC: Highland Hiker, 1996.

In Defense of Plants. "In Search of Stewartia ovata." Podcast, June 26, 2016. http://www.indefenseofplants.com/podcast/2016/6/26/ep-62-in-search-of-the-mountain-camellia-stewartia-ovata.

Jamerson, Bill. "Dollar-a-Day Boys: The Civilian Conservation Corps." Presentation and musical performance, Franklin, NC, March 12, 2016.

Jeffries, Stephanie, and Thomas R. Wentworth. *Exploring Southern Appalachian Forests: An Ecological Guide to 30 Great Hikes in the Carolinas, Georgia, Tennessee, and Virginia.* Chapel Hill: University of North Carolina Press, 2014.

Johnson, Becky. "Launching a Legacy: Nantahala Outdoor Center." *Smoky Mountain News*, June 6, 2012.

Jolley, Harley E. *"That Magnificent Army of Youth and Peace": The Civilian Conservation Corps in North Carolina, 1933–1942.* Raleigh: North Carolina Office of Archives and History, 2007.

Kasper, Andrew. "Balsam Lake High and Dry as Tourist Season Hits Full Stride." *Smoky Mountain News*, June 12, 2013.

Kauffman, Gary, National Forests in North Carolina botanist-ecologist. "Biological Assessment Managing Recreational Uses on the Upper Segment of the Chattooga Wild and Scenic River Corridor." December 5, 2011. http://www.fs.usda.gov/Internet/FSE_DOCUMENTS/stelprdb5351117.pdf.

Kays, Holly. "Wildfire Impacts Range from Barely There to Complete Char, but True Effects Remain to Be Seen." *Smoky Mountain News*, February 1, 2017.

Kearney, Lewis, retired USFS. Personal interview, meetings and e-mail conversations, 2016–17.

Kelly, Christine A., et al. "Crossing Structures Reconnect Federally Endangered Flying Squirrel Populations Divided for 20 Years by Road Barrier." *Wildlife Society Bulletin* 37, no. 2 (2013): 375–79.

Kennedy, Payson, co-founder of Nantahala Outdoor Center. Personal interview, fall of 2016.

Kesler, Sally, companion to A. Rufus Morgan. Personal interview at Nantahala Hiking Club Cabin, Cartoogechaye, NC, summer of 2016.

Lane, John. *Chattooga, Descending into the Myth of Deliverance River.* Athens: University of Georgia Press, 2004.

Lauritsen, Marv. *Ageless Giants: A 400-Year History of Joyce Kilmer Memorial Forest.* Rhinelander, WI: Jensmarie Publishing Company, 1998.

Lea, Bill, former USFS Interpretive Specialist. "White Oak Bottoms, Woods Camp: The School at White Oaks Bottoms and W.M. Ritter Company" / "Interviews with Logging Camp Residents" / "W.M. Ritter Lumber Company at Rainbow Springs & White Oak Bottoms" / "The Murder of a Good Man" / "Trail of Tears Field Notes" / "Tsali Notes." Unpublished documents, 2001–5.

Leatherwood, Bill, son of former supervisor at Mead Paper Mill. Phone interview, October 2015.

Ledford, Gerald, regional logging and train historian. Interview with author and e-mail conversations, winter of 2016–17.

Leverette, Will. *A History of Whitewater Paddling in Western North Carolina.* Charleston, SC: The History Press, 2008.

Lewis, James G. *The Forest Service and the Greatest Good: A Centennial History.* Durham, NC: Forest History Society, 2005.

Logan, Clay, retired USFS. Personal interview, Brasstown, NC, October 2016.

Lombard, Frances Baumgarner. *From the Hills of Home in Western North Carolina.* Highlands, NC: James Patrick Bryson, 2015.

Los Angeles Times. "Cocaine-Carrying Chutist Was Ex-Policeman, Lawyer." September 12, 1985.

Lyvers, Carl, and Tony Lyvers, former residents/caretakers at lodge at Buck Creek. Personal interview, Clay County, NC, November 2016.

Mainspring Conservation Trust. "Bringing Honor to Nikwasi." YouTube, April 15, 2016. https://www.youtube.com/watch?v=7datmOp-YxM.

Margulis, Abigail. "Man Charged with Setting Fires in Macon County. *Asheville Citizen-Times*, November 30, 2016.

———. "Wildland Fires Spike in Otherwise Typical Season." *Asheville Citizen-Times*, April 16, 2016.

Martin, Brent, regional director of the Wilderness Society. Personal interview, Sylva, NC, January 2017.

———. Guided hike. Wilderness Society and Southern Appalachian Plant Society. Overflow Creek on Bartram Trail, March 22, 2016.

Martin, Elizabeth Marie. "Holocene Environmental History of Panthertown Valley in the Blue Ridge Mountains of North Carolina." Western Carolina University, March 2014. http://libres.uncg.edu/ir/wcu/f/Martin2014.pdf.

Marx, Elizabeth. "Vegetative Dynamics of the Buck Creek Serpentine Barrens, Clay County, North Carolina." 2007. http://citeseerx.ist.psu.edu/viewdoc/download?doi=10.1.1.508.4404&rep=rep1&type=pdf.

McClung, Marshall. *Mountain People, Mountain Places.* Robbinsville, NC: Graham County Historical Society, 2006. Also see *More Mountain People, Mountain Places.*

McClung, Marshall, retired USFS. Personal interviews, meetings, guided hikes and e-mail conversations, 2015–17.

McConnell, Owen Link. *Unicoi Unity: A Natural History of the Unicoi and Snowbird Mountains and Their Plants, Fungi and Animals.* Bloomington, ID: AuthorHouse, 2014.

McIntosh, Gert. *Highlands, North Carolina: A Walk into the Past.* Highlands, NC: Gert McIntosh, 1983.

McRae, Barbara. *Franklin's Ancient Mound: Myth and History of Old Nikwasi.* Franklin, NC: Teresita Press, 1993.

———. "Preserving History at Wayah Bald." *Franklin Press*, August 13, 2010.

———. "Siler a Woman of Distinction in 19th Century." *Franklin Press*, April 8, 2016.

———. "Wayah Bald Landmark to Be Restored." *Franklin Press*, July 29, 2009.

McRae, Scott. "CCC Veterans Honored during Arrowood Glade Ceremony." *Franklin Press*, October 17, 1983.

———. "Despite Cuts, USFS Has Complete Program." *Franklin Press*, October 15, 1982.

———. "Forest Service Project Gives Wayah Bald Tower a New Look." *Franklin Press*, November 25, 1983.

———. "Special Ceremony to Mark 50th Anniversary of the CCC Planned." *Franklin Press*, September 21, 1983.

———. "Volunteers Play Big Role at U.S. Forest Service and LBJ Job Corps." *Franklin Press*, April 29, 1982.

Meacham, Jody. "Nantahala Gorge Accepted as Part of National Forest." *Asheville Citizen-Times*, June 22, 1973.

Message of the President of the United States Transmitting a Report of the Secretary of Agriculture in Relation to the Forests, Rivers, and Mountains of the Southern Appalachian Region. Senate Document 84. Washington, D.C.: Government Printing Office, 1902.

Miller, Lawrence, and Jeffrey DeBell. "Current Seed Orchard Techniques and Innovations." USDA Forest Service Proceedings, 2013.

Minard, Nancy. "Fun on the Wayehutta ATV Trail System." *Blue Ribbon* (March 2003).

Molloy, Johnny. *Hiking North Carolina's National Forests.* Chapel Hill: University of North Carolina Press, 2014.

Mooney, James. *Myths of the Cherokee and Sacred Formulas of the Cherokee.* Nashville, TN: Charles and Randy Elder, Booksellers and Publishers, 1982.

Moore, Carl S. *Clay County, Then and Now.* Franklin, NC: Genealogy Publishing Service, 2008.

Morgan, A. Rufus. *From Cabin to Cabin: The Life of A. Rufus Morgan.* Booklet. N.p.: self-published, 1980.

———. *History of St. John's Episcopal Church.* Pamphlet, Franklin, NC, 1974.

———. "The Nantahala National Forest." Chapter section in *Guide to the Appalachian Trail in the Southern Appalachians, Publication No. 8.* 4th ed. Washington, D.C.: Appalachian Trail Conference, 1960.

Morgan, Albert Rufus, III. Personal interview and tour. Franklin, NC, summer of 2016.

Morrison, Clarke. "Purchase Protecting Devil's Courthouse." *Asheville Citizen-Times*, March 15, 1990.

National Forest Reservation Commission. "Progress of Purchase of Eastern National Forests Under Act of March1, 1911 (The Weeks Law)." 1920. Walter Julius Damtoft Collection, D.H. Ramsey Library, UNC–Asheville. http://toto.lib.unca.edu/booklet/progress_purchase_eastern_nat_forests/default_progress.

Neal, Dan. "Following the Cherokee Footsteps Across WNC." April 11, 2016. http://www.citizen-times.com/story/news/local/2016/04/10/following-cherokee-footsteps-across-wnc/82465214.

Neal, Harry Edward. *The People's Giant: The Story of the TVA*. New York: Julian Messer, 1970.

Neely, Sharlotte. *Snowbird Cherokees: People of Persistence*. Athens: University of Georgia Press, 1991.

NeSmith, Eric. "Popular Yellow Mtn. Trail Re-Routed on Public Land." *Highlander*, June 8, 2001.

Newfont, Kathryn. *Blue Ridge Commons: Environmental Activism and Forest History in Western North Carolina*. Athens: University of Georgia Press, 2012.

Newman, Wanda, daughter of TVA dam worker. Personal interview at Fontana Marina, April 2016.

New York Times. "Cocaine and a Dead Bear." December 23, 1985.

North Carolina Bartram Trail Society. *NC Bartram Trail Newsletter* (Summer 1993).

———. *The 1993 Biennial Meeting of the Bartram Trail Conference*. Brochure, October 17–24, 1993.

———. "Take a Hike through History." http://www.NCBartramTrail.org.

North Carolina Department of Environment and Natural Resources. "Cullasaja River: Natural and Scenic River Feasibility Study and Recommendations." March 1999.

North Carolina Off-Highway Recreation. Cullowhee: North Carolina Off-Highway Vehicle Association Inc., 2001.

North Carolina Wildlife Resources Commission. "Balsam Lake Will Not Receive Trout Stockings This Season." March 27, 2011. http://www.Ncwildlife.org/News.

Nothstein, William. Interviewed by Dr. Louis Silveri. Transcript. Southern Highlands Research Center, University of North Carolina–Asheville, July 1, 1976.

Ogletree, William. "Making of a Radioman." Nantahala, North Carolina. http://www.nantahalanc.com/untitled8.html. Copied from http://www.jacksjoint.com/makings_of_a_radioman.htm.

Padgett-Atkisson, Joanna. "Access: Jackrabbit Mountain Bike and Hiking Trails." Unpublished document, no date.

Parce, Mead. *Railroad through the "Back of Beyond."* Hendersonville, NC: Harmon Den Press, 1997.

Parris, John. "Discussions Open Today on 190-Mile Extension to Blue Ridge Parkway." *Asheville Citizen-Times*, December 11, 1966.

———. "First Job Corpsmen Hit Franklin's Camp Arrowood." *Asheville Citizen-Times*, February 20, 1965.

———. "Nantahala National Forest Is Waterlogged This Year." *Asheville Citizen-Times*, March 13, 1975.

Paxton, Percy. "The National Forest and Purchase Units of Region Eight." USDA Forest Service, unpublished report, July 1, 1950.

Peattie, Roderick, ed. *The Great Smokies and the Blue Ridge: The Story of the Southern Appalachians.* New York, 1943.

Pittillo, Dan, retired Western Carolina University professor. Guided hike in Joyce Kilmer Memorial Forest, personal interview in the car en route to the hike and e-mail conversations, July 2016–January 2017.

Poole, Cary Franklin. *A History of Railroading in Western North Carolina.* Johnson City, TN: Overmountain Press, 1995.

Pratt, Joseph Hyde. "Corundum, Its Occurrence and Distribution in the United States." USDI National Park Service, 1906.

Radley, Laurel Cargill. "Francis Barr Cargill." *Episcopal Church Women, by Word and Example*, March 23, 2014. http://www.ecw-nc.org/by-word-and-example/?currentPage=3.

Raleigh News and Observer. "Parkway Bill Signed." October 11, 1968.

Rankin, Robert. "History of the Snowbird Mountain Lodge." Unpublished article, n.d.

Ransom, Todd. *Waterfalls of Panthertown Valley*. N.p.: self-published, 2013.

Ray, John R., and Malcolm J. Skove. *Bartram Trail: Detailed Guide with Maps of the Trails in Nantahala National Forest in North Carolina.* Central, SC: John Ray, 2005.

———. *Chunky Gal Trail and Fires Creek Rim Trail.* Central, SC: John Ray, 2010.

Reiche, Carl A. *Whither the Beloved Mountains: 26 Years of Hiking in and Fighting for the Westernmost NC Mountains*. Spiral-bound booklet. Green Mountain, NC: Carl Reiche, 1986.

Rhodarmer, Mia. "Cherohala Skyway History." 2008. http://www.cherohala.org/history.html.

Riggs, Brett. "Trail of Tears." Lecture. UNC–Asheville, Reuter's Auditorium, August 20, 2016.

Riggs, Brett, and Lance Greene. *The Cherokee Trail of Tears in North Carolina: An Inventory of Trail Resources in Cherokee, Clay, Graham, Macon and Swain Counties.* Report submitted to the National Park Service, Trail of Tears National Historic Trail, Santa Fe, New Mexico. University of North Carolina Research Laboratory of Archaeology and the Trail of Tears Association, North Carolina Chapter, 2006.

Ross, Cindy. "Savoring Tsali." *Smoky Mountain Living*, October 1, 2010. http://www.smliv.com/features/savoring-tsali.

Sakowski, Carolyn. *Touring the Western North Carolina Backroads*. Winston-Salem, NC: John F. Blair, 1990.

Sanders, Brad. *Guide to William Bartram's Travels*. Athens, GA: Fevertree Press, 2002.

Sandhusen, Hildegard. *Little Journeys*. Booklet. Franklin, NC: Teresita Press, 1996.

Sargent, Bill. "Standing Indian a Camper's Dream." *Asheville Citizen-Times*, March 21, 1983.

Sargent, Ralph M. *Biology in the Blue Ridge: Fifty Years of the Highlands Biological Station, 1927–1977*. Highlands, NC: Highlands Biological Foundation Inc., 1977.

Satterwhite, Bob. "Proposed ATV Trail Expansion Draws Criticisms." *Asheville Citizen-Times*, June 28, 1992.

———. "Wasilik Poplar Still Going Strong." *Asheville Citizen-Times*, August 22, 1993.

Scott, Bob. "Burned Tower to Get New Life." *Asheville Citizen-Times*, November 16, 1992.

———. "New Shelter Airlifted into Appalachian Trail Location." *Asheville Citizen-Times*, February 11, 1998.

Scott, Bob, mayor of Franklin. E-mail communications, December 2016.

Sexton, Ken. "Falcon Project Takes Flight Here." *Franklin Press*, July 24, 1989.

———. "Siler Bald Being Restored." *Asheville Citizen-Times*, May 21, 1990.

Shaffner, Randolph P. *Heart of the Blue Ridge: Highlands, North Carolina*. Highlands, NC: Faraway Publishing, 2001.

Shands, William E. "The Lands Nobody Wanted: The Legacy of the Eastern National Forests." Forest History. http://www.foresthistory.org/ASPNET/Policy/WeeksAct/LandsNobodyWanted_Shands.pdf.

Sharpe, Bill. *A New Geography of North Carolina*. Vols. 1–4. Raleigh, NC: Sharpe Publishing Company, 1958.

Shields, Johnny, president of the Smoky Mountain Off-Road Vehicle Club. Personal interview, January 2017.

Shireman, Douglas. "US Torpedo Troubles during World War II." History.net, February 1998.

Shope, Wade, son of Woodrow Shope, former timber cruiser for Nantahala National Forest. Personal interview at the Macon County Historical Museum, summer 2016.

Siler, Margaret R. *Cherokee Indian Lore and Smoky Mountain Stories*. Franklin, NC: Teresita Publishing, 1938.

Smith, Wally. "Ghosts of Panthertown." *Forest* (Winter 2010).

Smokey Mountain Off Road Vehicle Club 4 (Fall 1994).

Smokey Mountain Off Road Vehicle Club 6 (Spring 1995).

Smokey Mountain Off Road Vehicle Club 9 (Spring 1996).

Southern Appalachian Bicycle Association. "Jackrabbit Mountain Bike and Hiking Trail Project." October 1, 2016. http://www.sabacycling.com.

Spira, Tim. "Pollinators." Presentation. Asheville-Buncombe Technical College, March 8, 2016.

———. *Waterfalls and Wildflowers in the Southern Appalachians: Thirty Great Hikes.* Chapel Hill: University of North Carolina Press, 2015.

Stephanson, Mary. "Joyce Kilmer Memorial Forest." *Wildlife in North Carolina* 39, no. 9 (September 9, 1971): 4–6.

Stepp, Joe. "Notes on Aquonc." http://www.nantahalanc.com/NotesByJoeStepp.html.

Sutter, Karl. Guided history cruise on Fontana Lake, April 9, 2016.

Swank, Wayne T. USFS Coweeta Hydrologic Station. Phone interview, fall 2016.

Swank, Wayne T., and D.A. Crossley Jr., eds. *Forest Hydrology and Ecology at Coweeta.* New York: Springer-Verlag Inc., 1988.

Swank, Wayne T., and Jackson R. Webster, eds. *Long-Term Response of a Forest Watershed Ecosystem.* New York: Oxford University Press, 2014.

Sylva-Herald & Ruralite. "Envirnomental Concerns Part of Planning for ATV Course." August 15, 1991.

———. "Expansion Is Possible for Area ATV System." April 22, 1992.

———. "Ranger District Plans ATV Trail on Wayyehutta [*sic*]." September 13, 1990.

Tarrant, Michael, et al. "A Crowding-Based Model of Social Carrying Capacity: Application for Whitewater Boating Use." *Journal of Leisure Research* 28, no. 3 (1966): 155–68.

Taylor, David. "Plant of the Week: Quill Fameflower (*Phemeranthus teretifolius*)." USDA Forest Service. http://www.fs.fed.us/wildflowers/plant-of-the-week/phemeranthus_teretifolius.shtml.

Taylor, Robin, USFS seed orchard manager, Beech Creek Genetic Resources Management Area. E-mail communication, November–December 2016.

Taylor, Ron. *A Bear Paw Ramble.* Blairsville, GA: Straub Publishing, 2013.

Taylor, Ron, author. Personal interview, Bear Paw Village, Murphy, NC, fall of 2016.

Tennessee Valley Authority. "Hiwassee Dam." https://www.tva.gov/About-TVA/Our-History/Built-for-the-People/Torpedo-Testing-at-Hiwassee.

———. *Hiwassee Dam Project.* Booklet. Knoxville, TN: self-published, February 1938.

———. *Upper Little Tennessee River Region.* Knoxville, TN: Tri-State Development Association, June 1968.

Tinsley, Kathy. Personal interview at Nonah Weaver's Cabin, summer of 2016.

Trail Scout Chronicle 11 (Summer 1997).

Trail Scout Chronicle 12 (Spring 1998).

Trail Scout Chronicle 15 (Spring 2000).

Turner, Nancy. *The Summer Times: A Guide to Adventures around Highlands, Cashiers and Toxaway.* Tampa, FL: Cider Press Inc., 1994.

University of North Carolina–Chapel Hill. "The Unbuilt Blue Ridge Parkway: The Georgia Extension" (Fall 2013). http://projects.dhpress.org/unbuiltparkway/unbuilt-projects/ga-parkway-extension.

USDA Forest Service. *Carolina Connections.* Various articles, including "Get in the Water," "74-Year-Old Lookout Tower Gets Much Needed Repairs," "Take a Ride on the Jackrabbit Trails," "Explore the Backcountry," "Ride Your Horse on Forest Trails" and "OHV Trail Rules." 2011.

———. *Coweeta Hydrologic Laboratory: A Guide to the Research Program.* Booklet, October 2009.

———. "Establishment and Modification of National Forest Boundaries and National Grasslands: A Chronological Record, 1891–2012." *FS Bulletin-612.* 2012.

———. "Genetics Resource Management." http://www.fs.fed.us.

———. InciWeb-Incident Information System. New Release updates on WNC wildfires. USFS in conjunction with multiple agencies, November–December 2016.

———. Joint Information Center Posts to Provide Wildfire Information. Daily press release reports, November–December 2016.

———. *Nantahala National Forest.* Map, 1926.

———. *Nantahala National Forest: Cheoah and Tusquitee Ranger Districts.* Map, 2012.

———. *Nantahala National Forest: Early History of the Region.* Brochure, 1926.

———. *Nantahala National Forest: Georgia, North Carolina, South Carolina.* Booklet, 1936.

———. *Nantahala National Forest: Nantahala Ranger District.* Map, 2012.

———. "National Forests in North Carolina: March 2012 Briefing Book." Unpublished document. http://www.fs.usda.gov/Internet/FSE_DOCUMENTS/stelprdb5355450.pdf.

———. "Potential Species of Conservation Concern for the Nantahala and Pisgah NFs Plan Revision." Unpublished document, April 24, 2014. https://www.fs.usda.gov/Internet/FSE_DOCUMENTS/stelprd3797968.pdf.

———. "Rare Habitats on the Nantahala and Pisgah National Forests." Draft document, March 2014.

———. "Reforestation and Genetics." http://www.fs.fed/us.

———. *The Story of the Coweeta Hydrologic Lab*. Film, 2009.

———. "Wasilik Poplar." Letter to Dr. George Ammann from District Ranger Larry Hayden, December 19, 1986.

———. "Wasilik Poplar." Unpublished document, August 22, 1969.

———. "Wasilik Poplar Dedication." Spoken by USFS supervisor Peter Hanlon and USFS deputy supervisor H.C. Eriksson, August 31, 1969.

———. *Waters of Coweeta.* Booklet, 1953.

USDI National Park Service. *The Bartram Trail: National Scenic/Historic Trail Study*.

U.S. Fish and Wildlife Service. "Noonday Globe Snail." https://www.fws.gov/raleigh/species/es_noonday_snail.html.

Van Doran, Mark. *Travels of William Bartram.* New York: Dover Publications, 1955.

Van Noppen, Ina, and John J. Van Noppen. *Western North Carolina Since the Civil War.* Boone, NC: Appalachian Consortium Press, 1973.

Wager, Kit. "Thornton Dropped Drug before Jump, Agent Says." *Lexington Herald-Leader*, September 1985.

Welch, Ray, former member of CCC. Personal interview, Franklin, NC, March 16, 2016.

Wigginton, Eliot, ed. "Rev. A. Rufus Morgan." *Foxfire 4* (1977).

Wilcox, Leota Denton, descendant of John Denton. Personal interview and guided tour of the Denton Family Museum, Robbinsville, NC, summer of 2016.

Wilderness Society. *North Carolina's Mountain Treasures: The Unprotected Wildlands of the Pisgah and Nantahala National Forests.* Sylva, NC: self-published, 2002.

Wild South. "Discover Eastern Cherokee History." Cherokee Journey. http://cherokee.wildsouth.org/about.

Williams, Buzz. "Director's Page." [The fight over the headwaters of Chattooga River.] *Chattooga Conservancy* (Summer 2008).

WNC Hike Network. "Hiking the Panthertown Valley." 1996–2017. https://www.hikewnc.info/trailheads/panthertown-valley.

Young, Kelly V. "History of Fontana Village: The Village of Five Lives." Unpublished article in Fontana Village archives, n.d.

Zahner, Robert. *The Mountain at the End of the Trail.* Highlands, NC: Robert Zahner, 1994.

Zeigler, Wilbur G., and Ben S. Grosscup. *The Heart of the Alleghanies, or Western North Carolina.* Raleigh, NC: Alfred Williams & Company, 1923.

ABOUT THE AUTHOR

A retired nurse practitioner, Marci Spencer is the author of *Clingmans Dome: Highest Mountain in the Great Smokies* and *Pisgah National Forest: A History*, both published by The History Press. Her children's book, *Potluck, Message Delivered: "The Great Smoky Mountains Are Saved!,"* was published by Grateful Steps. The Yosemite Conservancy included Marci's essay "Pine Siskins Make History" in its book, *The Wonder of It All: 100 Stories from the National Park Service*, published to celebrate the centennial of the national park system.

Visit us at
www.historypress.net

This title is also available as an e-book